OUT OF YOUR
MIND

OUT OF YOUR MIND

THE LINKS BETWEEN BRAIN AND BODY

BRENDA TURNER

Copyright © 2010 Brenda Turner

The moral right of the author has been asserted.

Apart from any fair dealing for the purposes of research or private study, or criticism or review, as permitted under the Copyright, Designs and Patents Act 1988, this publication may only be reproduced, stored or transmitted, in any form or by any means, with the prior permission in writing of the publishers, or in the case of reprographic reproduction in accordance with the terms of licences issued by the Copyright Licensing Agency. Enquiries concerning reproduction outside those terms should be sent to the publishers.

Matador
5 Weir Road
Kibworth Beauchamp
Leicester LE8 0LQ, UK
Tel: (+44) 116 279 2299
Fax: (+44) 116 279 2277
Email: books@troubador.co.uk
Web: www.troubador.co.uk/matador

ISBN 978 1848764 231

British Library Cataloguing in Publication Data.
A catalogue record for this book is available from the British Library.

Typeset in 11pt Palatino by Troubador Publishing Ltd, Leicester, UK

Matador is an imprint of Troubador Publishing Ltd

Printed in Great Britain by the MPG Books Group, Bodmin and King's Lynn

For Barbara

CONTENTS

Chapter 1 – Atoms, Cells and Genes   1
What are atoms and molecules?
What are cells made of?
What are chromosomes?
What are genes?
What is DNA?
How do cells divide?
What are stem cells?
How do we evolve?
What is the selfish gene?
How do we inherit certain traits and features?
How does the brain and nervous system develop?
Disorders that are inherited
Disorders affecting cells and tissues

Chapter 2 – Blood (Cardiovascular) and Lymphatic Circulatory Systems   27
What does blood consist of?
Why are there different blood groups?
How does blood circulate round the body?
How does blood circulate in a foetus?
What is a pulse?
Which factors affect heart rate?
What makes the sound of a heart beat?
What is blood pressure?
How does blood clot?
Is the brain involved in the blood circulatory system?

Disorders affecting the circulatory System
What is lymph?
What function does the lymphatic system carry out?
Disorders affecting the lymphatic system

Chapter 3 – Brain and Mind   47
What are the main areas of the brain?
Does the brain really consist of grey matter?
How does the brain work?
Is the brain a sophisticated computer?
What and where exactly is the mind?
What is consciousness?
Disorders affecting the brain
Common psychiatric disorders

Chapter 4 – Digestive System   71
What happens to the food and drink we ingest?
Why are the kidneys so important?
How is the brain involved in digestion?
What makes us feel hungry?
What causes the stomach to rumble?
What makes us feel thirsty?
Why do we like or dislike certain foods?
Why do we need to smell food?
What makes us stop eating and drinking?
Why do some people under- or over-eat?
Which nutrients does the body need to thrive?
Why doesn't dieting work for some people?
Disorders affecting the digestive system
Disorders affecting the urinary system

Chapter 5 – Drugs and Addiction   99
How do drugs act on the brain and body?

Why do some people become addicted to drugs?
What are the effects of some common psychoactive drugs?

Chapter 6 - Emotions and Stress 115
What are emotions?
How many emotions do we have?
Why are facial expressions so important?
How do emotions develop?
Which areas of the brain are involved in emotional responses?
Why do we get angry?
Why do we cry?
Why do we laugh?
What makes us frightened?
Why do we blush when nervous or embarrassed?
What is stress?
Why does stress affect digestion?
How does Post-Traumatic Stress Disorder affect the body?

Chapter 7 – Hearing 139
What is sound?
How do we hear?
How does the ear transmit sound?
Can the unborn baby hear?
Why do we feel giddy and unbalanced when we twirl?
How does the brain convert air-waves into meaningful sounds?
How does the brain interpret music?
Why does too much noise make us feel ill?
What causes hearing loss?
What is ringing in the ears?
Why do some people 'see' sounds as colours?

Chapter 8 – Hormones (Endocrine System) 157
What are hormones?

What does the endocrine system consist of?
Which parts of the brain are involved in the endocrine system?
Disorders affecting the endocrine system

Chapter 9 – Language 173
How do we talk?
Which parts of the brain are involved in language?
Why can't other primates speak like humans?
How do we recognise the meanings of words?
Why do some people have difficulty learning to read and write?
Is there a language gene?
Is language instinctive?

Chapter 10 – Learning and Memory 193
How do we learn?
How do we remember?
Where and how are memories 'stored' in the brain?
Why do we forget?
How do we remember smells?
How can we improve our memory?

Chapter 11 – Movement 215
How do we move?
Which are the main bones of the body?
What does bone consist of?
Disorders affecting bones
What are the different types of joint?
What makes a joint crack?
Problems affecting joints
What are the different types of muscle?
How does muscle action work?
What makes a muscle tired?
What causes the knee-jerk response?

How does the brain control movement?
What causes Parkinson's and Huntington's diseases?
Other disorders affecting movement

Chapter 12 – Nervous System 237
What is the nervous system?
What does the central nervous system consist of?
What does the peripheral nervous system consist of?
How do nerve cells work?
What causes seizures?
What causes the sensation of 'pins-and-needles'?
Disorders affecting the nervous system

Chapter 13 – Pain and Touch 251
What is pain?
Why is pain good for us?
How does the sensation of pain reach the brain?
Why don't we feel pain in emergency situations?
What is referred pain?
What is phantom limb pain?
How do placebos work?
How do pain relief techniques work?
How do we experience different sensations of touch?
Which areas of the brain are involved with touch?

Chapter 14 – Respiratory System 267
Why must we breathe?
How does the respiratory system work?
Why are the lungs different sizes?
Why is there so much bleeding when the nose is damaged?
What is the function of nasal sinuses?
Is it true we can 'hear' through the nose?
What happens when we cough and sneeze?

How do we extract smells from the air?
Why are adenoids and tonsils more important for children?
What makes a new-born baby start breathing?
Is the brain involved in the respiratory system?
Why do we hyperventilate when stressed?
Why do we yawn?
Disorders affecting the respiratory tract

Chapter 15 – Sex and Reproduction   283
What are the mechanics?
Why do we need two sexes to reproduce?
How is the brain involved in sexual attraction?
What is the difference between heterosexual male and female attitudes to sex and reproduction?
Do humans have attracting pheromones?
How does Viagra® work?
Is there a gay gene?
How does alcohol or drugs affect sex and reproduction?
What is cloning?
Disorders affecting the reproductive system

Chapter 16 – Skin, Hair and Nails   305
What are the functions of the skin?
How is the brain involved with the skin?
What is the skin composed of?
What happens when skin is broken?
What makes skin different colours?
Why do we get goose pimples and shiver?
What causes dry and cracked skin?
Why do we sweat and why do feet smell?
What are fingerprints composed of?
What are blisters?
What is hair composed of?

Why does hair turn grey?
Disorders affecting skin and hair
What are nails made of?

Chapter 17 – Sleep and Dreams   323
Why do we sleep?
What happens when we sleep?
What is our biological clock?
Can we live without sleep?
Can we learn whilst asleep?
Which brain regions are involved in sleep?
What kind of sleep is caused by general anaesthetics?
Why do we dream?
Why do we have recurring dreams?
Why don't we remember our dreams?
Disorders affecting sleep

Chapter 18 – Taste and Smell   343
How do we discriminate between different tastes?
How does the brain receive taste information?
Why do we lose our sense of taste when we have a cold?
Are you a supertaster?
How do we discriminate between different smells?
How does the brain receive smell information?

Chapter 19 – Vision   355
How do patterns of light become images in the brain?
What causes light to become coloured?
What is colour-blindness?
Can we trust what we see – illusion and reality?
Why do we need to make eye-contact?
Does eating carrots improve eyesight?
Disorders affecting the eye

CHAPTER 1

ATOMS, CELLS and GENES – THE DEVELOPING BODY

What are atoms and molecules?
What are cells made of?
What are chromosomes?
What are genes?
What is DNA?
How do cells divide?
What are stem cells?
How do we evolve?
What is the selfish gene?
How do we inherit certain traits and features?
How does the brain and nervous system develop?
Disorders that are inherited
Disorders affecting cells and tissues

What are atoms and molecules?

An atom is a tiny particle. At its centre is a nucleus of protons and neutrons. Whizzing randomly around the nucleus are electrons in an outer shell. This movement is created by electrical forces. Protons have a positive charge, neutrons have no charge and electrons are negatively charged. Each atom has an equal number of positive and negative particles, resulting in a neutral status. Figure 1.1 shows a diagrammatic form of an atom.

Figure 1.1 The Atom

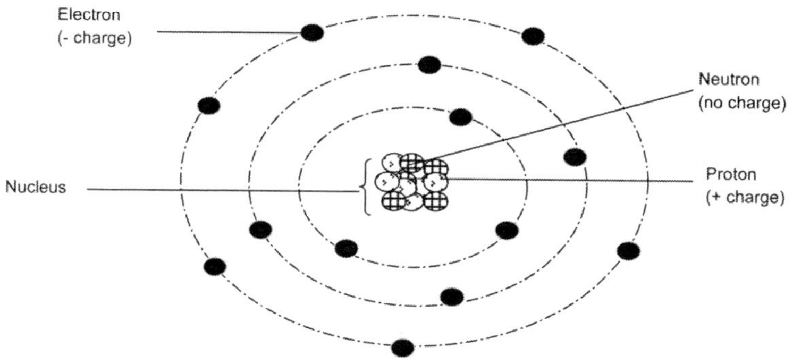

Electrons are about 1000 times smaller than protons and rotate around the nucleus at different levels of energy. Each level holds an increasing number of electrons. Atoms react with one another when the number of positively charged protons differs from the negatively charged electrons. When this happens there is a recombination to achieve a net result. In order to maintain stability, atoms must lose, gain or share their electrons with other atoms.

An element can have different atomic forms. This is because of the varying number of neutrons it contains. These different forms are called isotopes. For example the normal form of carbon is carbon-12 which has 6 neutrons in its nucleus. The isotope we may all recognise is carbon-14. It has 8 neutrons. It is used in radio dating to determine the age of once-living organisms.

A molecule is a mixture of two or more atoms. They can be of the same or different types of element. For example, the chemical compound for water is H_2O which contains two atoms of hydrogen and one atom of oxygen. Some molecules are closely bonded, as the elements share their electrons and retain their stability. Some molecules become more fragile when their electrons are shared, and will readily break up under certain circumstances, because of their weak bonds.

The human body breaks down some molecules such as sugars or fats, in order to release the energy it needs. For example, carbohydrates are sugar molecules consisting of carbon, oxygen and hydrogen. Glucose is a simple form of sugar used by cells in the body. It can be broken down to release water and carbon dioxide needed as fuel for energy. Lipids are molecules comprising carbon, hydrogen and oxygen atoms, and are hydrophobic, that is they do not mix with water. So they cluster together in large groups. Because of their weak bonding, lipids can easily change their shape. Press a 'fatty' part of your body and feel it move about. Lipids are fats which insulate and protect vulnerable areas of the body. Fats also store energy for emergency use.

Nucleotides are another important group of molecules. Adenosine triphosphate (ATP) contains ribose, adenine and three phosphate groups. When energy is released from the breakdown of other molecules such as sugars, the atom bonds are split and become atoms of adenosine diphosphate (ADP), with only two phosphate groups. ADP is an essential part of the process for making proteins, as will be discussed in the section relating to genes.

What are cells made of?

The human cell is the smallest part of the body and contains genetic material. Cells group together to form different tissue types which will shape the whole body. Figure 1.2 shows the structure of a typical cell.

Figure 1.2 Typical human cell

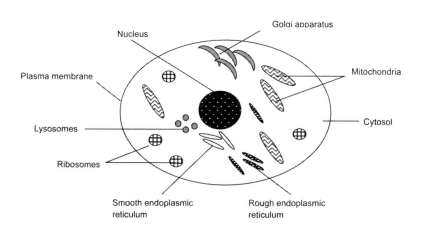

There are about 100 trillion cells in the human body. A single cell, called a zygote is the product of the combination of the female egg cell and the male sperm cell. The wall of the cell is called the plasma membrane and surrounds various little organs suspended in a fluid called cytosol, which is composed mainly of water and salts. The fluid itself contains a cellular scaffolding of protein strands which helps the cell maintain its shape. Tiny microfilaments also allow some cells, such as muscle to contract.

The plasma membrane consists of two layers of phospholipids, which are long chains of fatty acids. Embedded within these layers are special transport protein molecules, which control the movement of substances such as glucose across the membranes. The phospholipid molecules have an electrically charged hydrophilic head which allows water to interact, and a hydrophobic tail which repels water. The heads are on the outer surfaces of each layer and the tails face inwards. This is a very effective barrier preventing hydrophilic molecules passing freely in or out of the cell. Attached to some membrane proteins are branched carbohydrate molecules which partially immunise the cell against invaders. Figure 1.3 shows a diagram of plasma membrane.

The nucleus contains the genetic material of the cell, storing all the information needed for growth, metabolism and reproduction.

Mitochondria are shaped like sausages and are involved in aerobic respiration. They are often described as the cell's power house, as they generate the supply of ATP which is used as a source of chemical energy. Protein enzymes cause a chemical reaction between food molecules and oxygen. There can be

Figure 1.3 Plasma Membrane

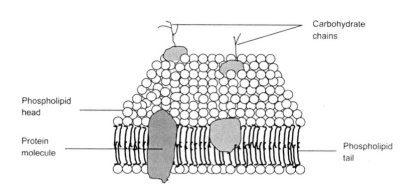

several thousand mitochondria in each cell. The cells that contain the most mitochondria, and are therefore most active, are in the liver, muscle and sperm.

Ribosomes are composed of ribonucleic acid (RNA) used to synthesise proteins.

The endoplasmic reticulum (ER) systems are interconnecting canals. The smooth type synthesises lipids and steroid hormones. In muscle cells it stores electrically charged calcium ions. Rough ER has ribosomes attached to its surface ready for the export of proteins from the cells to other parts of the body.

The Golgi apparatus consists of flat, folded membrane pouches. Proteins move from rough ER and are processed through granules that secrete them to the plasma membrane

Lysosomes are secretory granules which are produced in the

Golgi apparatus. They contain enzymes that digest worn out particles of molecules such as RNA, DNA and carbohydrates. The smaller units are then pushed out of the cell as waste material and transported to other parts of the body for removal.

What are chromosomes?

Chromosomes are threadlike molecules formed principally from proteins. They are composed of long pieces of DNA and replicate by mitosis (see page 13). Figure 1.4 shows the structure of a replicated chromosome. Humans have two types of chromosome: autosomes and gametes which are the sex chromosomes.

Figure 1.4 Replicated chromosome

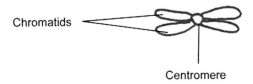

Chromatids

Centromere

Before replicating, the chromosome contains a single DNA molecule. After replication it consists of two paired chromatids, each with its own DNA molecule, joined to a centromere. Each molecule is like a parcel containing 46 chromosomes that fit into the nucleus of a cell. 44 are paired autosomes, that is there are 22 identical pairs. The remaining 2 are sex chromosomes. In females, there are 2 'X' chromosomes, and males have one 'X' from the mother and one 'Y' chromosome from the father.

What are genes?

Genes are chunks of inherited information that influence our development. There are 24,500 genes in the human genome.[1] They contain codes to build blocks of 20 amino acids into thousands of types of different protein molecules. Proteins form the basic structure of the human body. Some protein enzymes speed up or slow down chemical reactions. All the genes in the body are called its genotype and result from the fusion of genes from each parent. The genetic information is stored in the nucleus of each cell. Genes are stored in molecules of DNA. The genotype and its association with the environment are called the phenotype. A pair of genes is located in a particular part of the chromosome and there are different types of a gene called its allele. When more than one type is located in the same area, this affects the behaviour or appearance of an individual. For example, eye colour is determined by two different alleles. Some gene alleles have dominant and others recessive characteristics. Using the same example, the colour brown is more dominant than blue. So if one parent has brown eyes and the other blue, more of their children will have brown eyes than blue.

Genes can be switched on or off. Although each cell contains information which can relate to all parts of the body, only the switched on genes are active in a cell. For example the gene for a type of fat cell found in bone marrow is only switched on in parts of the bone. Elsewhere in the body the gene is switched off.

What is DNA?

The full name of DNA is deoxyribonucleic acid. It is a

complex molecule containing thousands of genes that carry genetic information. The way it transfers this information is marvellously complicated, using a four letter code to create thousands of different proteins. Each protein contains its own distinctive set of amino acids. DNA is a double-stranded helix like a twisted ladder. These two strands intertwine and interact through connecting 'rungs' of the ladder. It is the only molecule to self-replicate. At its centre is a fibre axis.

The strands are made up of chains of a sugar called deoxyribose, a phosphate group and a base. There are four different bases: adenine, guanine, cytosine and thymine and they can be described for simplicity by their initial letters, A, G, C and T. Two bases pair together on either side of the strands under a specific set of rules: T only pairs with A and C only pairs with G. The pairs sit along the vertical 'rungs' of the ladder, joined together by hydrogen bonds. The sugar and phosphate bonds on the upright strands of DNA are strong, whereas the hydrogen bonds are weak, and can be easily broken apart. When DNA replicates, one strand remains unchanged and the other synthesises into a new strand. Figure 1.5 illustrates the DNA molecule.

Genetic information is transferred by proteins. Coded instructions are passed from the nucleus of a cell to the cytoplasm in two separate steps, called transcription and translation. Transcription is carried out through an intermediary molecule called messenger ribonucleic acid or mRNA for short. It only has one strand. During this process, a portion of DNA unwinds, and the genetic information, through its base sequence is copied onto a new mRNA strand. mRNA also uses four bases, but substitutes the thymine for uracil (U).

Figure 1.5 DNA double helix molecule

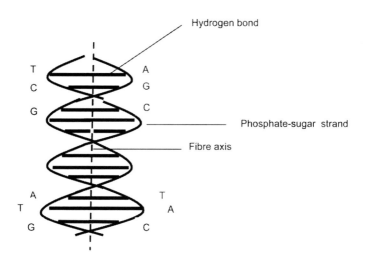

Also the two have different sugars, DNA uses deoxyribose and RNA has ribose. These two substances have slightly different structures.

Using one strand of DNA as a template, the weak hydrogen bonds between the base pairs break and the strands unwind. At a specific part of the DNA, the enzyme making the mRNA is given a start signal, and moves along the parent strand to pair up with the single base letter to form a new strand. So the DNA base letters A, G. C and T pair with U, C, G, and A respectively on the mRNA strand. When it reaches a stop signal, the new mRNA molecule is released from the nucleus, and the DNA double helix reforms. Figure 1.6 illustrates the process.

In the cell, the ribosomes translate the coded base pairs using the information on the mRNA to bind to another kind of RNA

Figure 1.6 DNA transcription

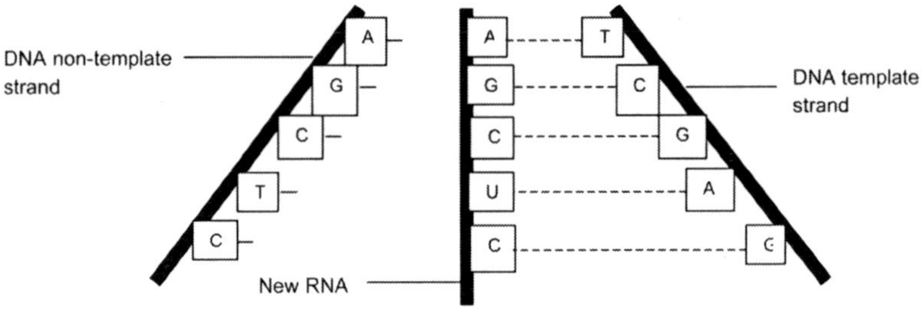

molecule called transfer RNA (tRNA), in sequences of groups of three bases, called codons. Each triplet codes for a specific amino acid. There are 64 codons including some that start and stop a sequence. As shown in Figure 1.7 the ribosome moves along the mRNA, and forms a protein from the amino acid codons. When the stop codon is reached, the new protein is complete. This is then used by the cell or exported to other parts of the body as required.

Figure 1.7 DNA translation

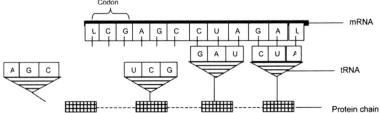

How do cells divide?

There are basically two methods: mitosis and meiosis. Figure 1.8 shows a comparison between the two processes. Each autosome (non-sex) chromosome inside a cell is duplicated from previous chromosomes by mitosis. The cell makes an identical copy and then divides in two with each new daughter cell having the full complement of 46 chromosomes. This happens in four phases, as illustrated in Figure 1.9.

Figure 1.8 Mitosis and Meiosis

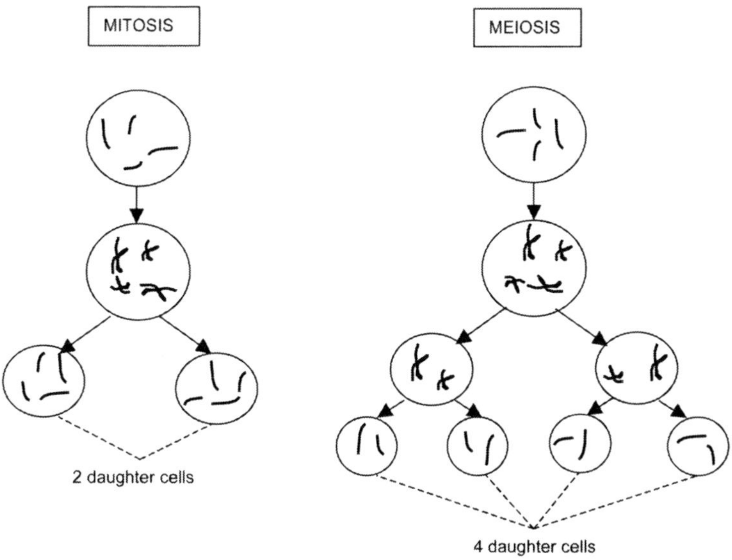

During the first prophase, the membrane surrounding the chromosomes disintegrates and the cell no longer has a nucleus. Very fine threads anchor at either end of the cell, and the other end attaches itself to the centromere at the centre of each double chromosome. During the next stage called

Figure 1.9 Cell division by mitosis

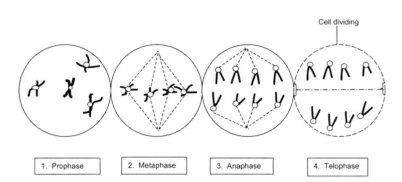

metaphase, these threads are pulled taut, aligned in the centre of the cell. At the third anaphase, the chromatids separate and one part of each pair migrates to either end of the cell, and the thread contracts. At the final telophase, the threads disappear leaving one set of chromosomes at each end of the cell. The chromosomes then uncoil and a new membrane surrounds each cluster of chromosomes forming a nucleus at each end. The cell then begins to divide in half to form two identical daughter cells.

Nuclear division by meiosis only occurs in the gamete (sex) cells, that is the reproductive female egg cell and male sperm cell. During this process, the cells duplicate twice to produce four daughter cells instead of the two as in mitosis. Each one is genetically different from the other and also from the parent cells, which means there is a unique combination for each new cell. When the egg and sperm gamete cells unite at fertilisation, they produce two different daughter cells. The chromosome pairs randomly transfer genetic information to either of the two new cells, which then divide again, and the new cells only contain one chromosome from each original pair, that is half the full complement. When the new cells fuse with other gamete cells the full number of chromosomes combine, and further divisions occur by mitosis (see also chapter on Sex and Reproduction).

What are stem cells?

Mammalian stem cells have the ability to renew themselves by dividing and differentiating into specialised types of cell. There are three main types: embryonic, adult and umbilical cord blood stem cells. In the developing embryo, the stem cells change into all the different types of tissue. In adults, stem cells repair and renew specialised cells. Stem cells can be grown and transformed into specialised tissue cells such as muscles or nerves. Because of their flexibility, there has been much scientific research into the possible advantages of using stem cells to repair or regenerate damaged organs, mostly carried out on mice. The first scientific reports were produced in 1963.[2]

There are four classes of stem cell. Totipotent stem cells are produced from a fertilised egg and can differentiate into embryonic and extra-embryonic types of cell. Pluripotent stem cells come from totipotent cells and can differentiate into any type of cell. Multipotent stem cells only produce cells from a closely related family of cells. Finally unipotent cells can only produce one type of self-renewing cell.

Embryonic stem cells are derived from the tissue of a blastocyst, which is an embryo about four of five days old, consisting of about one hundred cells. They develop into each of up to 200 types of different cells in adults.

Adult stem cells are able to divide and duplicate, and also create a cell that is more complicated. They are rare and can only produce cells within a range of related cells. For example, hematopoietic stem cells differentiate into the different types of blood cells.

Umbilical cord blood stem cells are from a fully developed umbilical cord and can be stored to treat blood disorders in children.

There is much controversy and continuing debate about the ethical use of human stem cells, especially embryonic. This is because the cells are obtained from embryos from fertility clinics which, although due to be destroyed, are still considered by some to be a human being, which should not be used in this way. Some adult stem cell therapies have been carried out such as bone marrow transplants used in the treatment of leukaemia.

How do we evolve?

An interesting fact is that about 90% of DNA has no known function. It is known as 'Junk DNA'. There are various explanations, one of which is that we have not yet discovered the full complexity of DNA. Another explanation is that it is a relic of our evolutionary past and no longer serves any useful purpose.

With processes as complicated as the transfer of genetic

material, there are of course mistakes which result in mutations. These errors in base pair sequences result in copying faults during cell division. The mutation transcribes to mRNA which translates into a faulty protein. This can lead to inherited diseases, such as cystic fibrosis or haemophilia. For example Queen Victoria produced a reproductive mutation which resulted in haemophilia in one of her sons and several later generations. Famously, her granddaughter Alexandra, the last Russian Tsarina, carried the disease to her son.

Mutations create variations in the genotype. Many are harmless and cause no adverse affect on our development. Harmful mutations are often slowly eliminated through the generations by natural selection. For example, if a mutation results in poor health, there is less likelihood of surviving to reproduce and pass it on to the next generation. Some can be positively advantageous under certain conditions, which means the genome is strengthened and the population evolves in a slightly different form. The genetic condition sickle cell disease alters the shape of red blood cells from round to a rigid crescent shape. This is caused by abnormal haemoglobin proteins sticking to the surface of the cells. This has a benefit in countries such as Sub-Saharan Africa, where malaria is rife, as the parasite cannot penetrate the altered structure of the cell, and spread the disease. Regrettably this is not an advantage in more temperate climates.

Over the generations, advantageous mutations gradually alter the human genotype. One result is that we have evolved to walk upright, leaving our hands free to carry out other tasks in the environment. Another example is that we have evolved opposable thumbs, which allow us to grip delicate objects and create tools. A mutation that allows the brain to learn more quickly will be copied to future generations, while the slower learners will gradually be overtaken.

Evolution by natural selection is a very slow process. Changes in our behaviour can take many generations. But there is another type of evolution which involves the environment and this has a much faster effect on our overall phenotype. It is called cultural evolution. Human beings are social animals and learn and copy from each other. Language is a rapid form of communication for sharing new ideas. Advances in technology and the ability to physically shape our environment to suit our needs has caused our brains to become larger and more structurally complex.

What is the selfish gene?

Richard Dawkins, the acclaimed evolutionary biologist coined the term in his book, first published in 1976.[3] What he meant by the term is that genes will maximise any opportunity to reproduce themselves in future generations. He describes organisms as survival machines that use genes for this and no other purpose. Successful genes will communicate their information, whilst unhealthy genes will not regenerate as the individual will not survive to pass them on. But there is no 'grand plan'. Mutations are spontaneous deviations and natural selection weeds out the good from the bad.

An example of the selfish gene would be the Black Widow spider which, after mating, has evolved to eagerly eat the sperm donor. Not good news for the males, but his job for life (literally) is complete, genes passed on, and the reproductive line continues. There are many other species that die soon after reproducing. So, what is good for the gene is not necessarily conducive to a long life.

How do we inherit certain traits and features?

Genes are paired on chromosomes, each part copied from a parent. The combination of paired chromosomes determines the trait or feature of an individual. Some genes are dominant and will be reproduced, whilst others are recessive and will only become active if present on both parts of the pair of genes.

Some inherited genes are sex linked. The Y chromosome is shorter than the X, which has more genetic material. For example the gene for colour vision is coded on the X chromosome. If a female inherits a faulty copy of the gene, she will still have perfect colour vision because the gene pair on her other X chromosome is normal. She can however carry the faulty gene on to a son, in which case he may be colour blind, because he only has the one faulty X chromosome, and his Y chromosome does not carry the gene. This is why colour blindness is more prevalent in males.

How does the brain and nervous system develop?

In a fully formed human brain there are about 100 billion neurons and 100 trillion interconnections. During the period of maximum growth of an embryo, each minute more than 250,000 neurons are added.[4]

The first human zygote cell begins to divide and subdivide

about twelve hours after conception. Within a week, the embryo has developed three separate layers, the ectoderm, endoderm and mesoderm.

Part of the ectoderm will cluster at the top end of the embryo to form the neural crest. This will migrate to become the neurons and glia cells of the peripheral nervous system. The ectoderm will also form the neural tube which will become the central nervous system. The inner part of the tube will form the ventricles of the brain, connecting to the central canal of the spinal cord. This layer will develop into the outer part of the body such as the skin, hair and nails.

The endoderm will form the gut, the inner lining of the digestive tract including the stomach, and the inner lining of the respiratory tract including the lungs. It also forms the glands including the liver and pancreas.

The mesoderm will form muscle, outer covering of internal organs, excretory system, the inner layer of skin, the ovaries and testes. Segments of mesoderm called somites will develop along the length of the embryo, and become the skeleton and the circulatory system, including the heart and blood vessels. The notochord is a flexible long structure of mesodermal cells which will eventually become the vertebral column or spine.

The layers thicken to form a flat oval shaped plate with a groove which will close up to become the neural tube. The interior of the tube will develop into the ventricular cavities of the brain and the central spinal cord. The plate will form the brain at the top end and the spinal cord at the bottom end. By the end of the third week from conception, the brain area develops separate swellings that will become the forebrain, midbrain and hindbrain. Figure 1.10 illustrates stages of development of the nervous system of an embryo at 28 days and 78 days.

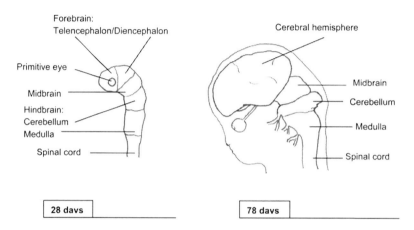

Figure 1.10 The developing nervous system

There are six developmental phases of the nervous system:

The first phase is called neurogenesis when stem cells of the neural tube keep on dividing and will eventually become neurons and glial cells. At the second phase, neurons migrate to set up separate communities that will eventually form different types of nerve cells at various locations in the nervous system.

At their new locations during the third phase, the cells begin to differentiate into specialised forms, each one influencing the development of neighbouring cells. The cells transcribe into new protein structures by switching on certain genes. Some undifferentiated stem cells will also start to specialise into a specific type of cell

The fourth phase is called synaptogenesis. Axons and dendrites begin to grow from the cells and connections are made with other neurons. Chemical signals direct growth cones at the axon tips along specific routes.

At the fifth phase, some neurons are no longer needed and will begin to die. A substance called nerve growth factor is produced by target cells and is taken into the cells that have established connections.

During the final sixth phase of development, there is a reorganisation of the synaptic connections. Those that are needed and used will be strengthened and others will become weaker and eventually stop functioning, presumably because their particular connections are no longer required.

The brain and nervous system depend on genes to influence the production of cellular proteins for ongoing development until maturity. There are external factors which in turn play their part in controlling gene activity, such as the level of nutrition the developing foetus receives from the mother at crucial stages of growth. Stressful and deprived environmental factors can also adversely affect a child's progress towards adulthood, and have consequences throughout life.

Disorders that are inherited

Down's Syndrome produces physical and mental abnormalities from birth. It is caused when three copies of chromosome 21 are inherited instead of the normal two. The condition results in a short stature and specific facial differences. Malformed neurons in some parts of the brain causes mild to moderate learning difficulties. A naive outlook often results in a particularly loving and cheerful personality.

Fragile X syndrome is caused by an unstable element at the end of the X chromosome. It produces unusual facial features with some mental retardation.

Haemophilia is caused by a fault on an X chromosome. It results in the blood being unable to clot successfully. Therefore, even comparatively minor injuries can lead to an abnormally high loss of blood. It is carried by females to their sons.

Huntington's disease usually appears in middle age. It is caused by an abnormal gene becoming dominant. It produces jerky uncoordinated body movements. There is also a tendency to form a more aggressive personality and an inability to empathise with others.

PKU(phenylketonuria) is caused by the absence of an enzyme that breaks down the amino acid called phenylalanine, which is found in food. It builds up in the body and if untreated can produce seizures and mental retardation. PKU can be easily controlled through careful diet.

Disorders affecting cells and tissues. Occasionally cell tissue grows abnormally and produces tumours. These can be benign or malignant. Benign tumours grow slowly and the cells are similar to normal. They are usually enclosed within a confined area and cause no problem. Malignant tumours multiply very quickly, are unlike the original cells and spread to other neighbouring areas or to different parts of the body through lymph or circulate in the bloodstream. Causes may be genetic or environmental. Radiation or toxic chemicals absorbed by the body, can result in damage to the DNA and cause malfunctions.

REFERENCES

[1] Waugh, A., and Grant, A., (2006) in *Ross and Wilson Anatomy and Physiology in Health and Illness*, Elsevier Limited, p.432

[2] Becker, A.J., McCulloch, E.A., Till, J.E. (1963) "Cytological demonstration of the clonal nature of spleen colonies derived from transplanted mouse marrow cells" *Nature* 197, pp 452-4 also Siminovitch, L., McCulloch, E.A., Till, J.E. (1963) "The distribution of colony-forming cells among spleen colonies" *Journal of Cellular and Comparative Physiology* 62, pp. 327-36

[3] Dawkins, R., (1976) (2nd Edition 1989) *The Selfish Gene*, Oxford University Press

[4] Rosenzweig, M.R., Breedlove, S.M., and Watson, N.V.,(2005) Biological Psychology, Fourth Edition, Sinauer Associates, Inc., p.183

CHAPTER 2

BLOOD (CARDIOVASCULAR) and LYMPHATIC CIRCULATORY SYSTEMS

What does blood consist of?
Why are there different blood groups?
How does blood circulate round the body?
How does blood circulate in a foetus?
What is a pulse?
Which factors affect heart rate?
What makes the sound of a heart beat?
What is blood pressure?
How does blood clot?
Is the brain involved in the blood circulatory system?
Disorders affecting the circulatory system
What is lymph?
What function does the lymphatic system carry out?
Disorders affecting the lymphatic system

What does blood consist of?

Blood is a fluid that carries nutrients and oxygen to cells throughout the body. It also carries waste products for removal from the body. About 55% of the content of blood is a pale yellowish fluid called plasma. Plasma fluid content is about 90% water and the rest is made up of proteins, nutrients from digested foods, mineral salts, hormones, gases and waste products.

The plasma protein molecules are too large to diffuse into the capillary blood vessels, and they create pressure which pushes the blood round the body. The main proteins are albumin, fibrogen and globulin, and they contribute to the thick quality of blood.

The remaining 45% of blood is made up of cells, and there are three types: red, also called erythrocytes; white, also called leucocytes and platelets, also called thrombocytes. Blood cells are formed in red bone marrow.

The red blood cells are biconcave discs without a nucleus. They transport mainly oxygen and a little carbon dioxide in the protein haemoglobin. Haemoglobin contains iron and when it binds to oxygen, the blood cells become red, as found in arteries. When oxygen is lacking, the cells have a blueish colour, as found in veins. The flexible shape of the red blood cells ensures a maximum area for the exchange of gases in the capillary blood vessels (see chapter on Respiratory System). They are active for approximately 120 days and are constantly being replaced with new cells.

White blood cells are actually colourless and contain a nucleus. They are larger than red blood cells and can change their shape. There are two main types: granulocytes and agranulocytes. Granulocytes, which contain granules in their cytoplasm, defend the body by responding to bacteria or viruses or anything else they consider as unwanted material. There are two kinds of agranulocytes, which do not contain granules: lymphocytes and monocytes. Lymphocytes are part of the lymphatic system and produce T-cell antibodies. Monocytes produce T-cells which engulf and destroy harmful bacteria. They also provoke allergic reactions.

Platelets are smaller than red blood cells, and also have no nucleus. They are made in red bone marrow. Their main function is to release chemicals that promote blood clotting and prevent further blood loss when a blood vessel is ruptured.

Why are there different blood groups?

An antigen is a substance that stimulates the production of an antibody in blood plasma, and responds to the action of any substance it considers foreign to the body. People inherit different antigens on the surface of their red blood cells. These determine the blood groups ABO and Rhesus. The ABO groups are A, B, AB, and O. The rhesus factor is another antigen, and a person can be positive or negative in any of the groups, which means there are eight different blood groups. Table 2.1 sets out the groups and their compatibility status, which is essential for transfusion purposes. In explanation, as you can see from the table, a donor with blood group A is incompatible with B and O because they both make anti-A antibodies which will react against the A antigens. Conversely, a recipient with blood group A is incompatible with B and AB because it makes anti-B antibodies.

Table 2.1 The ABO blood group system

Blood Group	Antigen	Antibody	Donor		Recipient	
			Compatibility	Incompatibility	Compatibility	Incompatibility
A	A	B	A and AB	B and O	A and O	B and AB
B	B	A	B and AB	A and O	B and O	A and AB
AB	A and B	none	AB	A, B and O	All groups	none
O	none	A and B	All groups	none	O A, AB and B	

The majority of people have the Rhesus positive (Rh⁺) antigen and do not make anti-Rhesus antibodies. Rhesus negative (Rh⁻) individuals can make anti-Rhesus antibodies. Parents with differing Rh factors can negatively influence the foetal blood supply, although this is usually easily rectified.

How does blood circulate round the body?

Blood moves around the body through blood vessels called arteries, veins and capillaries. The heart acts as a pump keeping the blood moving always in the same direction. Figure 2.1 gives a general outline of the routes taken by the blood vessels.

Figure 2.1 Blood circulation system

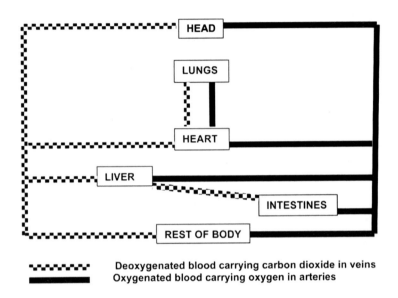

Arteries have thick walls consisting of three layers of tissue. The outer layer is fibrous, the middle layer has smooth muscle and elastic tissue and the inner layer has scaly tissue. They always carry blood away from the heart. The largest arteries which are near the heart have more elastic tissue that expands to meet the pressure demands of the blood. Smaller arteries are more muscular to cope with the pressure of blood flowing from the rest of the body. As they get farther away from the heart, arteries subdivide into smaller vessels called arterioles. They are also muscular which allows them to constrict and dilate as required.

Veins have thin walls and return blood to the heart. They have the same three layers of tissue. Some veins have valves which open to allow the blood to pass, but then close between heart beats to prevent the blood flowing backwards. There are more valves in areas of the body farthest from the heart, such as the lower limbs. They also subdivide into smaller vessels called venules.

Blood circulation begins and ends with tiny capillaries, which are the final sub-divisions of arteries and veins. They are single cell vessels which form networks and allow an exchange of gases and nutrients to permeate through their thin walls. Once the exchanges have taken place, the blood re-circulates. There are two blood vessel systems: pulmonary and systemic. Pulmonary circulation travels through the right side of the heart, which pumps the blood into the lungs, where there is an exchange of oxygen and carbon dioxide (see chapter on Respiratory System). Systemic circulation travels through the left side of the heart. This blood is pumped to the rest of the body, where waste material is excreted and nutrients and oxygen are picked up along the way.

The heart is a roughly oval shaped hollow organ, lying left of centre between the lungs, with the bottom end resting on the diaphragm. It has three layers. The outer layer called the pericardium is composed of two sacs containing a fluid which prevents friction when the heart beats. The middle layer, called myocardium has a unique cardiac muscle which contracts involuntarily. The thin inner layer is called the endocardium which has a smooth wall to allow the blood to flow easily.

The heart is divided into four sections. The top halves are the right and left atria and the bottom halves are the right and left ventricles. The largest veins in the body are the superior (above) and inferior (below) venae cavae. They enter the right atrium and the blood is pumped into the right ventricle and then into the pulmonary artery. It is the only artery to receive deoxygenated blood. This divides into left and right pulmonary arteries carrying the blood through the system to the lungs where oxygen is picked up and carbon dioxide passes out of the vessels.

Two pulmonary veins from each lung carry oxygenated blood back to the left atrium. The pulmonary veins are the only veins to carry oxygenated blood. The blood then passes into the left ventricle and is pumped into the aorta, which is the largest artery in the system. Various valves operate at each stage of the process to ensure the blood keeps flowing in a circular direction. Figure 2.2 shows the main parts of the heart and the direction of blood flow.

Figure 2.2 Anatomy of the heart

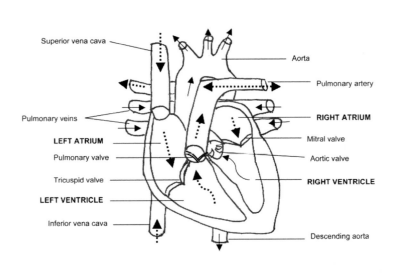

BLOOD AND LYMPHATIC SYSTEMS

The function of the heart is to maintain constant circulation around the body, through a pumping action. The heart contracts and relaxes in cycles of about 60 – 80 every minute. Cardiac output is the amount of blood leaving the heart at each contraction. In a healthy adult at rest, the output is about 70 ml per minute. This can increase dramatically when exercising, and an athlete can pump out up to 35 litres every minute.[1] Literally all 'hands' are to the pump, as the body utilises its reserves of oxygen and energy, to meet the increased demands from muscles.

How does the blood circulate in a foetus?

During its development in the womb the foetus lies in amniotic fluid and does not use the lungs. Oxygen and nutrients are obtained from the mother via the placenta and umbilical cord. The umbilical vein carries blood from the placenta. Half enters the inferior vena cava and the rest enters the liver, where it meets the portal vein and moves to the right atrium. There is an opening called the foramen ovale between the left and right atria and most of the blood flows through into the left atrium, bypassing the pulmonary circulation. It then flows into the left ventricle, and is pumped through the aorta to the rest of the body. Some blood re-enters the placenta which extracts carbon dioxide and waste material to be expelled through the maternal circulation system. The remaining blood from the right atrium enters the right ventricle and is pumped into the pulmonary artery. A small vessel called the ductus arteriosus connects the pulmonary artery and the aorta and directs the blood away from the lungs.

When the baby takes its first breath after birth, pressure

changes from the air filling the lungs, closes the foramen ovale and separates the two atria and the blood circulates through its two systems in the normal way. The ductus arteriosus also closes off within a few hours of birth.

What is a pulse?

The pulse is a wave of pressure caused by the contraction of the left ventricle which passes along the blood vessels. Figure 2.3 shows the main pulse points at arteries around the body.

Figure 2.3 Main pulse points

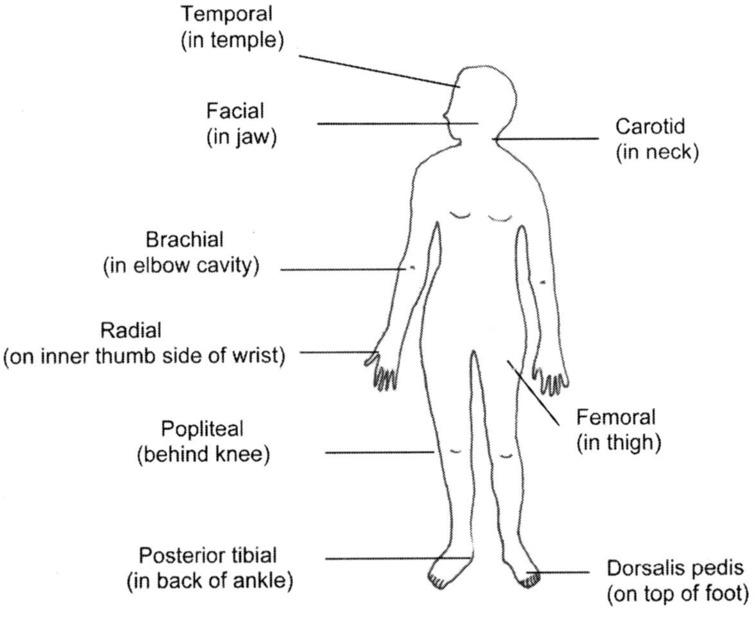

- Temporal (in temple)
- Facial (in jaw)
- Carotid (in neck)
- Brachial (in elbow cavity)
- Radial (on inner thumb side of wrist)
- Popliteal (behind knee)
- Femoral (in thigh)
- Posterior tibial (in back of ankle)
- Dorsalis pedis (on top of foot)

When the ventricle contracts, the blood is forced through the aorta, which stretches and produces the pulse wave. It can be felt by pressing the artery near the surface of the skin against the bone. An adult at rest has normally between 60 – 80 pulses per minute and can increase to 180-200 beats per minute during strenuous exercise. The pulse is a measure of heart rate. Pulse rates become slower with age. The pulse should also be evenly spaced, any irregularity indicating a potential heart or circulation problem.

What factors affect heart rate?

The rate of the heart beat depends on a variety of factors. The heart rate is faster in women than men, and the young have the fastest rates. Heart rate increases when body temperature rises, and also when exercising, to meet the extra demands of the muscles for more blood. The heart beats faster to cope with the effects of gravitation when a person is upright, than when lying down. Hormones and drugs circulating in the system can also increase heart rate, as well as high levels of carbon dioxide. The autonomic nervous system controls the heart rate, passing signals through the sympathetic and parasympathetic nerves.

What makes the sounds of a heart beat?

The heart acts as a double pump, through its pulmonary and systemic systems. There are normally two *'be- boom'* heart sounds (often described as a 'lub-dup'). The first *'be'* sound is created when the valves between the atria and the ventricles snap shut. The pressure in the ventricles rises as they contract.

This causes the blood to be pushed back towards the atria, and closes the valves which stop the backflow. The second *'boom'* sound is caused when the valves close between the aorta and the pulmonary arteries. As the ventricles empty their contents, the pressure falls and the blood attempts to reverse back causing the valves between them to close.

What is blood pressure?

Blood pressure is the force it exerts as it flows, on the walls of blood vessels. It is always measured in millimetres of mercury (mmHg). This is because the instrument used to measure it is a mercury-filled tube called a sphygmomanometer. In recent years this is often replaced by the use of a digital machine. Blood pressure is measured in two parts. Systolic pressure is when the ventricles contract, and diastolic pressure is when they relax. Normal blood pressure for adults is 120 mmHg systolic and 80 mmHg diastolic. The systolic pressure is the peak pressure in the arteries at the start of the cardiac cycle. Diastolic pressure is the lowest or resting phase.

Blood pressure varies at different times of the day, is higher in women than men and tends to become higher (hypertension) with increasing age. It is established by the volume of blood passing through and the resistance in the walls of the blood vessels. As we get older, the walls tend to stiffen, which causes more resistance and an increased pressure.

How does blood clot?

Blood clotting minimises loss from an injured blood vessel.

Platelets release an enzyme that alters the protein structure in blood plasma, which becomes thicker and stronger and acts as a protective shield over the area. The platelets then contract and produce a clear sticky serum. This helps to pull together the damaged blood vessel which then dries to form a scab. An allergic reaction causes an increased blood flow to the site and can be seen as a reddening of the skin around the wound. Another enzyme then begins the healing process and the scab slowly falls away as the repair is completed.

A thrombosis or blood clot inside an intact blood vessel can occur under abnormal circumstances. This is discussed further in the section relating to disorders affecting the circulatory system.

Is the brain involved in the blood circulatory system?

The cardiovascular centre is situated in the medulla and pons in the brainstem at the base of the brain (see chapter on Brain and Mind). Blood pressure is controlled. Signals are received from areas of the brain involved in emotions such as fear, anxiety or pain. The hypothalamus controls body temperature. In both cases changes in blood pressure and heart rate are effected by dilating or constricting blood vessels.

Baroreceptors are nerve endings which detect the pressure of the blood flowing through the vessels and send signals to the central nervous system to activate the sympathetic and parasympathetic nerves. This adjusts heart rate and alters the size of blood vessels to regulate the pressure to meet the needs of the body.

Disorders affecting the circulatory system

Acquired Immune Deficiency Syndrome (AIDS) is caused by the Human Immune Deficiency (HIV) virus. It attacks the T-white cells and makes them unable to defend the body against disease.

Anaemia results from insufficient quantities of haemoglobin to bind with oxygen to meet the needs of the body. The red bone marrow does not produce enough red blood cells. The main reasons for this are excessive blood loss due to internal or external injury, or a lack of iron in the diet. Women who are pregnant or menstruating are more susceptible, as are children whose swift growth needs to be sustained by greater oxygen levels.

Arteriosclerosis occurs when the artery walls become less flexible. It mainly affects larger arteries which develop fibrous tissue and calcium on the walls, and results in high blood pressure (hypertension).

Gangrene results from a lack of blood supply to the fingers and toes, which will decay.

Haemophilia is a condition whereby the blood is unable to clot resulting in excessive bleeding. Women with a mutated gene, although unaffected themselves, pass it on to male children who inherit the disease.

Heart failure can occur when the heart cannot meet the demands of the body. The left ventricle, which pumps out the largest amount of blood is the section of the heart usually affected. Sometimes the body compensates for problems by enlarging the cardiac muscles to make the ventricle thicker.

Hypotension is low blood pressure which means insufficient blood is supplied to the brain. It can result in fainting. This is not to be confused with **hypertension**, which is high blood pressure.

Leukemia is the result of the overproduction of white blood cells, which take the place of red blood cells in bone marrow, which in turn causes anaemia.

Thrombosis is caused by a thrombus or clot forming in the blood vessels and obstructing the free flow of blood. Coronary thrombosis occurs when a large artery becomes blocked and can cause a heart attack. Deep vein thrombosis (DVT) occurs when a thrombus forms in a vein of a lower limb.

Varicose veins are produced when the valves in the veins are faulty causing the blood to collect instead of freely circulate. They are found mainly in the lower limbs and can become swollen and painful. Haemorrhoids or piles are produced when varicose veins in the rectum or anus are put under pressure. This can be caused by constipation or in the late stages of pregnancy.

What is lymph?

Lymph is a clear fluid which is the same as the interstitial fluid which bathes tissue cells. Oxygen and nutrients leak from blood capillaries and drain into lymph vessels. The lymph transports them back into the blood stream, and also collects waste material and foreign substances such as viruses and bacteria, which are then destroyed.

What function does the lymphatic system carry out?

Figure 2.4 shows the main structures that make up the lymphatic system. One of its functions is to drain fluid from the tissues back into the bloodstream. The lymphatic system also acts as a filter for the blood and lymph. It plays a fundamental role in the immune system in defending the body against infection.

Lymph capillaries have more permeable walls than blood capillaries, allowing free transfer of material. They merge into larger vessels which contain valves to keep the lymph flowing in one direction. The muscular walls contract regularly to keep the lymph circulating and skeletal muscle movement near the lymph vessels also helps to move it along. The lymph vessels eventually form two large ducts, the lymphatic and thoracic ducts. The lymphatic duct is about 1 centimetre long, opening into the right subclavian vein, and drains lymph from the right side of the thorax, head, neck and right arm. The thoracic duct is about 40 centimetres long, opening into the left subclavian vein and drains lymph from the rest of the body.

At various strategic points the lymph vessels pass through glands called lymph nodes. The outer layer is fibrous tissue and the remaining tissue consists of macrophage cells, lymphocytes and lymphatic tissue. Macrophages destroy antigens by filtering the lymph as it passes through and the lymphocytes produce antibodies. The main organs connected to the lymphatic system are the spleen, the thymus gland, the tonsils and adenoids, the appendix and the lacteals in the small intestine.

Fig 2.4 The lymphatic system

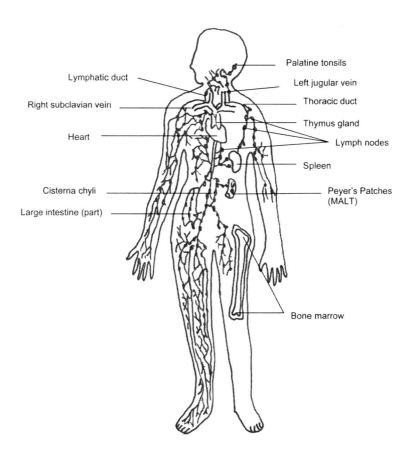

The spleen is like a large lymph node in structure. Macrophage cells remove and break down old and damaged red blood cells. The spleen also acts as a reservoir for blood which can be distributed to other parts of the body as required.

The thymus gland stores mature T-lymphocytes which only react to one type of antigen. They enter the blood stream as necessary. The thymus gland begins to decline after puberty and its effects deteriorate.

The tonsils and adenoids are located behind the nose and soft palate near the entrance of the breathing passages. They filter out some of the harmful substances we breathe in or swallow.

The appendix also contains lymphoid cells which should help fight infection. However, its removal appears to have no adverse affects on the body, and it is mainly considered to be a vestigial organ, that is it probably had a more useful function in our evolutionary history.

Mucosa-associated lymphoid tissue (MALT) is found in lacteals, which are lymphatic vessels conveying chyle from the intestines to the thoracic duct. Fats are absorbed in the small intestine and as a consequence, the lymph turns milky and is called chyle. MALT contains B- and T-lymphocytes from bone marrow and thymus. Peyer's Patches are collections of lymphoid follicles in the small intestine. The tonsils also contain this type of tissue.

Disorders affecting the lymphatic system

If the lymphatic system does not completely destroy tumours or infections, tiny fragments can settle in the lymph nodes where they accumulate and spread.

Glandular fever is an infectious virus which spreads to lymphoid tissue. It can begin as tonsillitis.

Hodgkin's or **non-Hodgkin's lymphomas** are malignant tumours of lymphoid tissue.

Lymphadenitis is an infection of the lymph nodes caused by microbes.

Lymphoedema is a swelling in a vessel caused by an obstruction, most commonly a tumour. The affected lymph nodes are surgically removed to prevent further growth.

The **spleen** can become enlarged. This is usually caused by the spread of other blood circulatory infections, such as typhoid fever, malaria or tuberculosis.

REFERENCES

[1] Waugh, A., and Grant, A., (2006), *Ross and Wilson Anatomy and Physiology in Health and Illness*, Tenth Edition, Elsevier, p. 87

CHAPTER 3

BRAIN AND MIND

What are the main areas of the brain?
Does the brain really consist of grey matter?
How does the brain work?
Is the brain a sophisticated computer?
What and where exactly is the mind?
What is consciousness?
Disorders affecting the brain
Common psychiatric disorders

What are the main areas of the brain?

Vertebrate brains are divided into three main regions: the forebrain, the midbrain and the hindbrain – see Figure 3.1.

The forebrain contains the diencephalon, which acts as an access point for information received from the spinal cord to the telencephalon. It helps to regulate the internal systems connected with the senses and movement. The upper part of the diencephalon is called the thalamus, and receives sensory information which is transmitted to the overlying cortex.

Figure 3.1 Main regions of brain

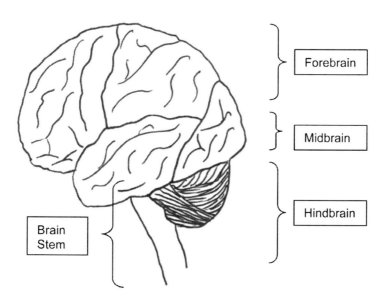

The lower part is called the hypothalamus, which although small contains nuclei receptive to vital bodily functions connected with hunger, thirst and keeping the body at a regular temperature. It also controls the pituitary gland, which is responsible for secreting hormones.

The telencephalon consists of a pair of large cerebral hemispheres. The outer layer of these is a highly folded surface area rather like a walnut, called the cerebral cortex. Figure 3.2 shows the division into four lobes, frontal, temporal, occipital and parietal.

Figure 3.2 Brain Lobes

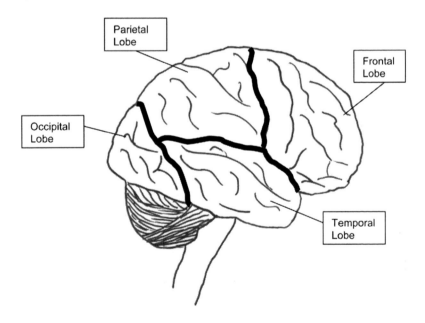

The frontal lobe area is connected with thinking, planning, organising and solving problems. It is also where emotions and behaviour are controlled.

The parietal lobe area controls our perception of the world, and our sense of space within it. Information received from the senses is integrated in this region.

The occipital lobe is primarily concerned with processing visual information. Most of the visual cortex lies within this area.

The temporal lobe processes memory and language and information received about the sounds we hear.

Figure 3.3 Separate functions of each hemisphere

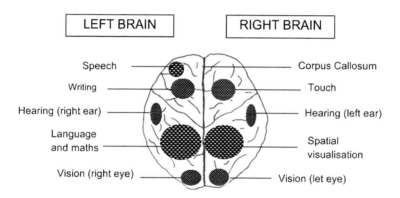

The cerebral hemispheres to the right and left of the forebrain are connected in the middle by the corpus callosum. This is a broad band of nerve fibres, which act as a communicator between the two hemispheres. Each hemisphere is responsible for different functions. The left side controls language, speech and writing, maths, logical and analytical thought. The right side controls touch and spatial visualisation. It is involved in creative processing, including art, music, intuitive and abstract concepts. Figure 3.3 shows a top view (superior) of the brain with the areas highlighted.

The midbrain (also known as the mesancephalon), contains two pairs of bumps. The top pair, called the superior colliculi controls the way we adjust to visual information and starts the processes of movement in response to these stimuli. The bottom pair, called the inferior colliculi controls information relating to sound. Both of these structures relay information to the thalamus, which then transmits it on to the cerebral cortex. Deep inside the midbrain is an organ called the substantia nigra, which is involved in the release of the hormone dopamine. It is believed that a reduction in the supply of dopamine is the main cause of Parkinson's disease.

The reticular formation is a long bundle of neurons stretching from the thalamus to the medulla in the hindbrain. One of its functions is to control sleep and arousal patterns.

The hindbrain receives auditory information about vibration and direction of sound. It consists of the pons, medulla and a cauliflower-like structure called the cerebellum.

The pons is a protruding structure that contains nuclei which relays and analyses information relating to movement, which is received from the senses. For example information transmitted from the ear enters the pons through the vestibulocochlear nerve (see chapter on Hearing). The information is relayed between the cerebral cortex and the cerebellum.

The medulla, lying between the pons and the spinal cord controls autonomic functions. It transmits signals between the brain and the spinal cord, and regulates breathing, heart rate and blood pressure. It also contains nuclei that are connected with taste, hearing, balance and the control of head and neck muscles.

Figure 3.4 shows approximate positions of these major structures in the brain.

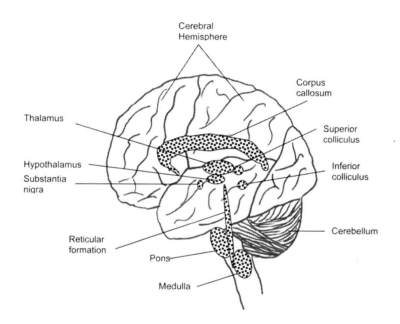

Figure 3.4 Brain structures

The cerebellum is, from an evolutionary aspect, the most primitive part of the brain. The name means 'little brain' in Latin. Similar structures are found in so-called lower class animals, such as fish, reptiles and birds (the rest of the brain is known as the neo cortex, which is highly developed in higher classification of living things such as mammals). The cerebellum lies over the brainstem (mid- and hindbrains). Like the cerebral cortex, it consists of a pair of highly folded structures, which are divided into lobes. However, unlike the cerebral cortex, the folds are tightly folded, accordion-like, in parallel grooves. The two halves are so closely joined they are treated as one organ.

The cerebellum plays an important role in the control and co-ordination of body movement. Recent imaging techniques have shown that it is also involved in language and learning skills. It reacts and processes information very quickly, and is involved in major decisions such as the best approach when facing danger.

Does the brain really consist of grey matter?

Figure 3.5 Cross section of brain showing Grey and White matter

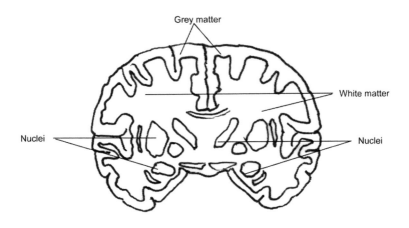

The grey matter of the brain is the external layer of the cerebral hemispheres. It is also found in interior parts of the brain in specialised areas, such as the thalamus and hypothalamus

amongst others, and also in the cerebellum. The areas comprise tightly packed neuronal bodies called nuclei, which receive signals from motor and sensory stimuli. These are processed and routed to the central nervous system. Because the information is analysed in this part of the brain, there is a popular conception that the 'grey cells' are where 'intelligence' lies.

The rest of the brain is composed of white matter, which consists of myelinated nerve cell axons (tails of nerves). These transmit information to the nerve cell bodies in the grey matter areas (see Figure 3.6). The axons have a myelin sheath that has a whitish appearance. Myelin is fatty material that electrically insulates the axons. This has the effect of speeding up the signals travelling along them. The sheaths are bundled layers of myelin, and at each end is a gap called a Node of Ranvier. The electrical signal – called an action potential, jumps from gap to gap, thereby speeding up transmission.

Figure 3.6 Myelinated nerve axon

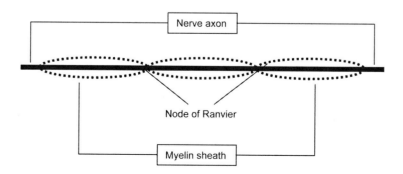

Myelinated axons are also found in the spinal cord transmitting information along it. Some degenerative diseases such as multiple sclerosis are caused when these axons become de-myelinated, because of malfunction of the particular cells that form myelin.

So the white matter transmits information to grey matter which analyses it and decides the best pathways for rerouting.

How does the brain work?

The human brain has more than a hundred billion neurons with trillions of connections.[1]

The human cortex contains areas attending to all aspects of behaviour. The primary motor cortex transmits information to the nerves dealing with movement in the spinal cord. The primary sensory cortex receives information directly via the thalamus, except for the sense of smell, which is received directly from the nose. All the areas are connected by nerve axons.

Information is also stored in the cerebral cortex as memories, and these pathways connect to areas that are directly involved in emotional states of mind. These are the basal ganglia, hippocampus and amygdala. The basal ganglia are clusters of nuclei linking the thalamus and the cerebral cortex. These are responsible for changing our movements to ensure actions are appropriate for the situation. If the organs in the basal ganglia are damaged, this can produce the decreased movements of Parkinson's disease or the uncontrolled movements in Huntington's disease.

There are many pathways that are activated by our behaviour. For example, if you touch an object, skin receptors respond to the temperature and feel of the object. If the object is too hot or too cold, tissue damage results and produces the feelings of pain or discomfort.

The brain uses 20% of the oxygen needed for the whole body. The blood circulates to provide the necessary nutrients, such as glucose and oxygen and removes carbon dioxide and other waste products. Exchanges of these essential nutrients and gases occur in tiny blood vessels called capillaries. The main blood vessels supplying the brain are called the carotid and vertebra arteries. Disruption of the blood flow to these arteries, caused by the blood clotting or damage, can result in stroke, blindness or paralysis.

The blood vessels in the brain are highly specialised, and will only allow very small molecules to pass through their membranes. This is called the blood-brain barrier, and tightly controls chemical changes to prevent unnecessary signals being passed along the nerve axons. The barrier also prevents many infections in the bloodstream from passing to the brain.

The brain and the spinal cord have cavities collectively called the ventricular system, filled with cerebrospinal fluid (CSF). Three protective skins or membranes surround the brain and spinal cord called the meninges, and it is the middle layer that contains the CSF. It flows in a circular path from four ventricles in the brain to the central canal in the spinal cord. It is through this fluid that harmful elements and essential hormones pass out of and into the bloodstream. The fluid also acts as a cushion allowing the brain to float safely within it.

Inflammation of these membranes caused by bacterial infection can lead to meningitis, which can result in permanent brain damage and can be fatal.

The brain needs fuel to work. This is supplied by elements contained in food such as carbohydrates and oxygen. When these elements mix, carbon dioxide, water and heat is generated. Some heat is used to maintain a consistent body temperature. The rest is stored as energy to be used as required in the form of glucose and oxygen.

Is the brain a sophisticated computer?

Both use electrical signals to transmit information
But, the brain also uses other methods. These are by chemical messengers called neurotransmitters. It is essential that both systems work in tandem. A computer works on a switch on/switch off binary basis and produces faster results. Information is processed serially, that is one action follows the next. The brain however, uses parallel processing, which means several actions can be carried out simultaneously. It is much more complicated. Interaction between chemical and electrical stimuli can result in excitation or inhibition at the neuronal synapses and alters the passage of the electrical signal (action potential) accordingly.

Both have a memory
A computer's memory is contained in its microchip. It can only be increased by the addition of more power or binary digits. The brain increases its own memories by adding more neuronal connections throughout its normal span of development, based on life experiences.

Both use energy as fuel
But the computer only needs a constant electrical supply to fuel its programs. The brain requires complex sets of interactions of gases (such as oxygen) and chemical minerals and vitamins (such as glucose) to meet its needs.

Both can adapt
But the computer can only be adapted, by adding to or removing existing software or hardware. The brain is continually learning new skills or adapting to meet changing internal or external conditions. It also has flexibility in analysing information, whereas a computer must be provided with an exact instruction. For example, if I type the word 'coyp', it would not recognise this as 'copy'. If I read the word 'coyp' , my brain would presume it had been misspelt and – depending on context – would correct and react accordingly.

The brain only needs partial information to understand the whole picture, based on its experience. For example, if I read 'the red ap... was juicy and crunchy to eat', my brain would comprehend the subject of the sentence was probably an apple.

Both get viruses
Intervention can sometimes correct and rebalance both the computer and the brain. But a computer must get new parts or be restored to its pre-virus state. In many instances, the brain can utilise alternative pathways to replace the damaged routes.

The Turing Test

The mathematician, Alan Turing devised a thought experiment test to answer the question: Can a machine show intelligence?[2] The test involved a human assessor asking questions to two participants who were hidden behind a screen. One participant was a computer, and the other another human. If the assessor cannot decide whether an answer came from the computer or the human, then it must be assumed the computer – later called the Turing Machine – was showing intelligence.

The main argument against this premise is that whereas a computer program has a set of rules, allowing it to perform, which could be similar to the processes used by the brain, only the brain can understand the meaning behind the action. You cannot get a sensible answer from a computer to the question: how are you feeling today?

The Chinese Room argument
The philosopher John Searle posed another famous thought experiment, this time showing that a computer cannot behave like the brain, because it cannot compute semantically.[3] A person who understands English, but not Chinese, is placed in a room with a book of rules in English and a large number of Chinese symbols. The rulebook shows how to match these symbols by their shapes. The person has no idea what the symbols mean, or what difference matching them can make to their meaning.

People outside the room can understand Chinese, and pass bundles of symbols which actually pose questions. Using the rule book, inside the room the person will match the symbols as instructed and pass them back. The matched symbols answer the questions posed. The person is acting exactly like a digital computer obeying the rules of the program without understanding the meaning behind the instructions.

The philosopher Daniel Dennet believes Searle's thought

experiment is too simplistic.[4] Strings of symbols being processed as a set of rules cannot produce understanding, no matter how complex the program.

Finally, a computer is not creative. It can never come up with new ideas, or alternative formats that have not been previously programmed into its system. Neither is it conscious or self-aware.

So the answer to the question seems to be a definite no – based on current computer knowledge. However, this may change in the future, as advances in technology increase the capacity of artificial intelligence. There is another train of thought that says if the complexity of a system keeps increasing, an emergent property will appear[5]. An analogy could be if two parts of oxygen are added to one part of hydrogen, the result is water, changing from a gaseous to a liquid state.

What and where exactly is the mind?

Throughout the centuries it has been considered that the mind is a separate entity to the brain, and this is called dualism. One of the most famous theories is that of the seventeenth century philosopher and mathematician René Descartes who believed the mind resided in the pea-shaped pineal gland at the centre of the brain between the two hemispheres.[6] He reasoned that most structures of the brain are duplicated, like the cerebral hemispheres. As the pineal gland was one organ, this is where the mind must be. (We now know that the gland releases a hormone called melatonin, which regulates seasonal changes in animals and also changes in puberty). It was believed that the mind consisted of a substance that was fundamentally different to physical material, which behaves according to the rules of physics and chemistry. The mind was 'psychological' and includes beliefs, emotions and desires. Descartes saw nature as a complex machine and human nature as a smaller machine within it. It has been called the ghost in the machine[7]. When challenged to clarify how the two kinds of substance interacted, he said that we had not yet reached a level of understanding to explain it. A refinement of dualism is the belief that minds and bodies never interact, but run synchronised parallel systems.

Monism is the view that all substances are material matter, and can interact accordingly. The mind is what the brain does. In an analogy to the computer, the brain is the hardware to the software of the mind. The computational theory of mind is one of cause and effect, and states that beliefs, emotions and desires can be processed to produce certain behaviours. However, as discussed above, unlike a computer it is not locked into circumscribed parameters. It is much more flexible and complex. Although we have not reached a level of scientific understanding showing exactly how this happens, monists believe the processes associated with the mind are explainable in material form and are allied to physical attributes.

What is consciousness?

Trying to explain consciousness is like holding soft jelly – it slips and slides about and changes its form. Consciousness is a subjective state of awareness. Being aware is considered to be a function of higher-order animals only. Therefore, there must be an evolutionary advantage. It enables us to choose, solve problems, decide and plan for the future.

BRAIN AND MIND

Consciousness performs certain functions. Being conscious allows us to function well, and adapt and learn from new and important events in our lives. We prioritise events by controlling our behaviour accordingly. Decisions are made to achieve our pre-set goals. We are able to define information received and place it into its correct context. We are able to change our views if we consider our previous intentions or beliefs are incorrect. We can also obtain information from unconscious sources if we consider it to be of current use.

The spiritual approach involves altered states of consciousness via religious experience or trance states. These are created by twirling, fasting or taking psychoactive drugs. There is the belief in the ability to consciously enter a parallel world. Some philosophers suppose it is not physical, but just ideas. Others believe that all matter is conscious, even pebbles on the seashore and the sea itself.

Physical theories try to explain behaviour by identifying phenomenal consciousness with actions in the brain. Studies with patients in a persistent vegetative state show damage to neural connections between the brainstem and the thalamus and higher cortical areas[8].

Following critical illness, some people describe what is known as 'out-of-body' experiences which appear to have remarkable similarities. They appear to be floating in a dream-like state, and can see themselves as separate entities, and can describe the furniture and the people in the room. This seems to show that consciousness is separate from the physical brain. However, controlled experiments have shown that the same experiences can be triggered artificially by stimulating certain areas of the brain. This means there may be physiological explanations for the phenomenon.

Each person's consciousness is unique – thoughts, feelings and desires. Conscious states have a qualitative character, which are called qualia (singular is quale). For example, the quale of a particular wavelength of electromagnetic radiation from the Sun is a colour. Qualia states are caused by neurobiological processes in the brain[9].

Disorders affecting the brain

Cerebral Palsy is a condition affecting movement. It is mainly caused by a lack of oxygen to a baby's brain just before, during or just after birth. Rarely, the cause can be by maternal infection during pregnancy or post-natal infection or injury. Movements are abnormal and there is difficulty with co-ordination and balance. Other symptoms may include defective hearing, seizures and some brain function impairment.

Encephalitis is an inflammation of the brain caused by a viral infection. The symptoms vary from mild to life threatening, and include headaches, weakness, problems with speech and hearing, and loss of consciousness. It can develop from glandular fever or complications arising from mumps or measles.

Epilepsy is a condition which causes seizures at irregular intervals. It can be caused as a result of a brain injury or infection, or birth trauma. There are different types called grand mal and petit mal depending on the seriousness of the symptoms. Petit mal usually occurs in children, and can cause temporary unconsciousness, sometimes several times a day.

The condition is usually outgrown. The causes are not known. Another type of epilepsy creates partial seizures, which are caused by abnormal electrical activity in the temporal lobe of the brain. The symptoms vary in complexity and can include unusual movements and disruption to the senses of smell, vision and taste.

Meningitis is an inflammation of the membranes covering the brain and spinal cord (meninges). It is caused by viral or bacterial infection. Viral infections are comparatively mild and usually occur during the winter. The symptoms are similar to influenza. The bacterial infections are rarer and can be life-threatening. The main symptoms are high fever, headache, nausea and vomiting. Other symptoms are a preference for a dim light and stiff neck. In many cases, there is also a rash under the skin which will not fade when pressed. Sleepiness and loss of consciousness may follow. Intravenous antibiotics are used to treat the condition, and vaccines are available for some types of meningitis.

Migraine is a severe headache which can last up to three days. Migraine with aura produces flashing coloured zigzag patterns of light in an area to one side of the face. There may also be tingling sensations or numb feelings. It last about an hour and can be followed by a headache. Migraine without aura produces a severe pain, usually to one side of the forehead, but sometimes at the back of the head. It is accompanied by feelings of nausea and a strong adverse reaction to light and sound. Cluster headaches can recur over periods of two to three weeks and affect one side of the forehead. The cause may be related to blood vessels in the brain becoming dilated and allowing the blood to circulate faster. Migraine appears to run in some families.

Strokes occur when the blood supply to the brain is disrupted. This causes the brain tissue to decay. It occurs mainly in the elderly, and can be exacerbated by hypertension (high blood pressure), smoking or diabetes mellitus. The symptoms are paralysis of one side of the body or a limb and an inability to speak or see clearly. Some strokes are triggered following an injury to the brain, causing haemorrhaging of the area. Sometimes a blockage of an artery can cause a temporary stroke called a transient ischaemic attack (TIA), which usually disappears within 24 hours.

Common Psychiatric disorders

Alzheimer's Disease causes nerve cells (neurons) in the brain to degenerate, and the brain to shrink in size. It mainly occurs in people over 60. Although the primary cause is not known, people with the condition have been shown to have an unusually large build-up of two kinds of protein. Plaques are deposits of beta-amyloid protein, which accumulate in the spaces between neurons. Tangles are deposits of tau protein which accumulates inside the neurons. The disease is progressive, as the deposits develop. At the initial stage, the patient becomes increasingly forgetful and short-term memory is particularly impaired. This causes confusion and disorientation and leads to anxiety and depression. Cognitive thought gradually becomes more dysfunctional and there can be personality changes causing verbal and physical violent behaviour. It is thought that life-style, family history, and age are all factors affecting the likelihood of contracting the disease.

Bipolar disorder also known as manic depression, causes episodes of extreme mood swings between high excitability and depression. It may be caused by chemical imbalances or abnormal electrical circuits in the brain, but there is no conclusive evidence so far. Antibiotic and antipsychotic drugs are used to treat the condition.

Dementia is usually caused by Alzheimer's disease. However, it can also be caused by cerebrovascular disease created by blocked or narrow arteries in the brain. This reduces the blood supply to the brain and causes progressive deterioration in cognitive function. It is most usual in people over 65, but can occur in younger people with brain injury or disease or alcoholism. The symptoms are similar to Alzheimer's disease.

Depression is a debilitating disorder. Patients have negative feelings of loss, sadness, lack of appetite and self-confidence, and reduced emotional responses. They often suffer from insomnia and are unable to function normally. Although there may be an obvious cause such as bereavement, in many cases the trigger cannot be defined. There may be a genetic factor as depression tends to run in families. Some mothers have antenatal depression, which can last for up to a year without medication. Anti-depressant drugs or psychotherapy can treat the condition.

Schizophrenia is a psychotic illness that usually occurs in early adulthood. Patients have irrational thought processes and disturbed emotional responses. Behaviour can be erratic and abnormal. The schizophrenic may also hear voices or feel controlled by dangerous influences. Sometimes multiple personalities can be expressed. It may be triggered by stress or genetic influences. As the disorder progresses, the patient slowly withdraws from human society and lacks motivation. Brain imaging scans have shown abnormalities in some areas of the brain. The neurotransmitter dopamine is also thought to be involved. The brain may become too sensitive to its effects or there may be an over-production. Dopamine pathways are connected to areas of the brain involved with emotions, memory and social behaviour. People who regularly use cannabis, cocaine or amphetamines may also show symptoms of schizophrenia. The condition may be temporary or may recur throughout life.

REFERENCES

[1] Pinker, S., (2002), The Blank Slate, BCA, p. 197
[2] Turing, A., (1950) 'Computing machinery and intelligence' *Mind*, **59**, pp.433-60
[3] Searle, J.R., (1990), 'Is the Brain's mind a Computer Program? *Scientific American* **262**, pp. 20-5 and 26-31
[4] Dennett, D.C., (1991) *Consciousness Explained*, Penguin Group, pp. 438
[5] Trefil, J., (1997) *Are We Unique?* Ch.13. Wiley
[6] Descartes, R., (1641), *Meditations on First Philosophy*, Paris
[7] Ryle, G., (1949), *The Concept of Mind*, Chicago, University of Chicago Press
[8] MacIver, M.B., Mandema, J.W., Stamski, D.R., Bland, B.H., (1996) Thiopental uncouples hippocampal and cortical synchronized electroencephalographic activity. *Anesthesiology* 84 (6), pp 1411 – 24
[9] Searle, J.R., (1992), *The Rediscovery of the Mind*, MIT Press

CHAPTER 4

THE DIGESTIVE SYSTEM

What happens to the food and drink we ingest?
Why are the kidneys so important?
How is the brain involved in digestion?
What makes us feel hungry?
What causes the stomach to rumble?
What makes us feel thirsty?
Why do we like or dislike certain foods?
Why do we need to smell food?
What makes us stop eating and drinking?
Why do some people under- or over-eat?
Which nutrients does the body need to thrive?
Why doesn't dieting work for some people?
Disorders affecting the digestive system
Disorders affecting the urinary system

THE DIGESTIVE SYSTEM

What happens to the food and drink we ingest?

Figure 4.1 Digestive system

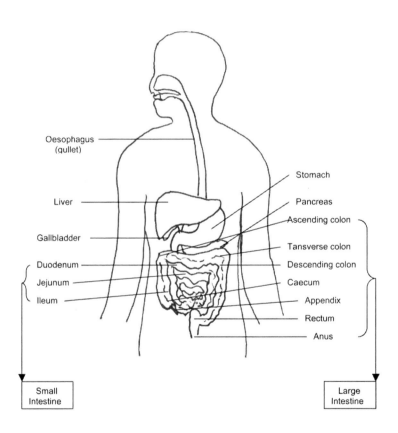

The digestive system is a series of hollow organs running from the mouth to the anus. The organs are lined with mucous containing glands that produce gastric juices and help digest food. Figure 4.1 shows the main organs responsible for digestion.

The food and drink we ingest needs to be changed into smaller molecules. These can then be absorbed into the bloodstream and carried to cells throughout the body, to meet its needs for energy and nutrition. The human diet needs carbohydrates, proteins and fats to provide the vitamins and amino acids necessary to build up proteins.

The cells use fuel in the form of minerals and vitamins and ions. Proteins are synthesised and ions transported across the walls of the cell. Energy in the form of heat, water and carbon dioxide are produced and released from the cell. The rate at which fuel is used by the body is called its metabolic rate. Figure 4.2 shows the movement of material to and from the cell.

Figure 4.2 Movement of nutrients in a cell

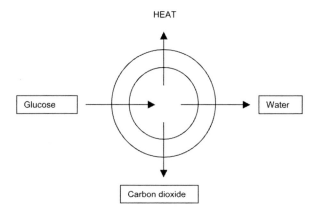

THE DIGESTIVE SYSTEM

Glucose is essential as fuel to be converted to energy. Any remaining glucose is chemically altered and stored. Some is converted to adipose fat in the liver and is transported and deposited to various sites throughout the body. The rest is converted to glycogen and stored in the liver and muscles.

Food moves through the system by muscular movements of the organ walls. The movement is called peristalsis and acts in a wavelike motion by narrowing and pushing food and fluid along the whole length of the tract.

When food enters the mouth, the teeth bite and grind it into a small ball, which mixes with saliva and is then swallowed. It passes through the oesophagus (or gullet), and when it reaches the stomach, which is like a large sac, a valve opens allowing the food to enter. This is then stored during the digestive process.

The wall of the stomach relaxes and expands to enable it to hold all the food and drink it receives. It contains glands that secrete the hormone gastrin which produces hydrochloric acid dissolving some foods. Gastrin is also important for the growth of the lining of the stomach, the small intestine and the colon. The stomach wall also secretes an enzyme called pepsin, which breaks down proteins into their constituent amino acids. As the food is broken down, it enters the small intestine where further chemical actions take place, using secretions received from the pancreas and the liver. The small intestine is over six metres long, and is divided into the duodenum, the jejunum and the ileum.

The pancreas sends enzymes to the duodenum, the first part of the small intestine. It also manufactures insulin, which enters the bloodstream and metabolises sugars.

The duodenum secretes a hormone called cholecystokinin (CCK), which controls the rate at which the stomach empties, by stimulating the digestion of fat and proteins.

The liver produces a digestive juice called bile that is stored in the gallbladder. Bile ducts release the chemical into the duodenum, and break down fats into small absorbable pieces. CCK causes the release of digestive enzymes from the pancreas and bile from the gallbladder. Other intestinal enzymes include maltase, lactase and sucrase, which process sugars.

Most of the digested molecules of food and water are absorbed through the walls of the small intestine and transferred into the bloodstream for use as storage, or for further chemical reactions throughout the body. The remainder is waste material and is emptied into the caecum, the first section of the large intestine. The waste is stored in the rectum and evacuated as faeces through the anus. Liquid waste is produced by the kidneys in the urinary system (see next section).

At the end of the caecum, near the junction of the small and large intestines is an organ about 10 cm long – the appendix. Its function is not known, although it does contain lymphoid cells which play a role in fighting infection. It can become inflamed and enlarged, causing the condition called appendicitis. It can be removed with no adverse affects. The usual explanation, in the place of any other, is that it is a remnant of the evolutionary process, and serves no current useful purpose to humans.

Why are the kidneys so important?

The urinary system comprises 2 kidneys, 2 ureters, the urinary bladder and the urethra. Figure 4.3 shows the structure of the system.

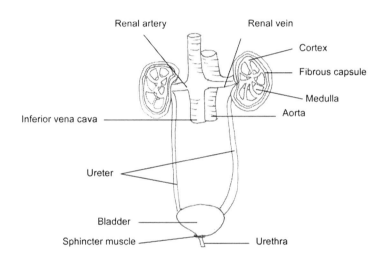

Figure 4.3 Urinary System

The kidneys perform three functions: they produce and secrete urine, which removes waste material from the body; they produce and secrete the hormone erythropoietin, which controls the formation of red blood cells, and they produce and secrete the enzyme renin, which controls blood pressure.

The kidneys are each shaped like a bean, approximately 11 cm long, 6 cm wide, 3 cm thick and weigh 150 g. They are situated on each side of the spine at the back of the abdominal cavity just below the diaphragm. Each one consists of an outer cortex and inner medulla and is surrounded by a fibrous capsule. The cortex contains specialised capillaries called glomeruli with a series of tiny tubules which together make up the nephrons. Nephrons have one open and one closed end. The closed end forms an indented glomerular capsule called Bowman's capsule, and encloses arterial capillaries. They filter blood under pressure and reabsorb water and other substances back into the blood. There are about 1 million nephrons in each kidney. The substances that are not reabsorbed make up urine. The tubules carry the urine to the renal pelvis and on through the ureter tubes to the bladder. In this way the kidneys regulate the fluid balance of the body by conserving it when required (for example after strenuous exercise when the body loses water in sweat), and excreting excess water.

Urine formed in the kidney passes through the papilla of the medulla and into the ureter. The smooth muscle walls of the ureter propel the urine into the bladder by peristalsis, in a similar way to the balls of food passing through the oesophagus and into the stomach.

The kidneys filter, reabsorb and secrete molecules during the production of urine. It is composed of 96% water, 2% urea and 2% of other substances that the body does not need, including ammonia, phosphates, sulphates, sodium, potassium, and chloride. It is coloured yellow, because it contains a bile pigment called urobilin. The glomerulus and glomerular capsule have semi-permeable walls, which allow water and other small molecules to filter through, whilst larger molecules including blood cells and plasma protein are unable to pass through the walls as they are too large. The kidneys produce 180 litres of filtrate every day, most of which is reabsorbed. Only about 1 litre is excreted as urine[1].

Material that is needed by the body is reabsorbed to maintain the required fluid balance, and also the correct level of pH (acidity/alkalinity) of the blood. All the glucose and amino acids are reabsorbed.

The bladder acts as a reservoir for urine prior to evacuation from the body. The walls stretch when there is a build up of over 300 ml. This stimulates nerve fibres. In babies, this causes a spinal reflex and urine is passed out of the bladder (micturation). When the nervous system has developed, the reflex sends signals to the brain indicating a full bladder, which makes us aware of the need to pass urine. Sphincter muscles in the urethra contract and will prevent this happening until convenient – for a limited time only!

How is the brain involved in digestion?

Feeding control is caused by the interaction of the hypothalamus and other areas in the forebrain and the brainstem. The hypothalamus has neuron receptors which are sensitive to glucose levels. Glucose is the only fuel used by the brain. Glucose receptors are also found in other areas of the body such as the liver.

Four peptide hormones, leptin, insulin, ghrelin and PYY produce signals in the brain that notify us when we should eat or drink. Leptin is secreted by fat cells and informs the brain about current energy stores. It helps to inhibit appetite. As previously mentioned, insulin is produced by the pancreas, and breaks down glucose into glycogen. It is released when the brain responds to the sight, smell or taste of food in anticipation of glucose arriving in the bloodstream. Ghrelin is secreted by the stomach and stimulates eating, whilst PYY secreted by the intestines, inhibits eating. These two hormones respond rapidly to changing hunger signals.

An area of the brainstem called the nucleus of the solitary tract (NST) integrates signals from the hypothalamus and also from inputs from nerves in the spine and cranial nerves in the brain. The liver communicates levels of glucose and fatty acids to the NST. The duodenum and the stomach also send signals to this area regarding nutrient requirements.

Parts of an area above the brainstem called the basal ganglia are also involved in assessing the pleasure obtained from eating and drinking.

What makes us feel hungry?

The human body manufactures some of the nutrients needed to supply energy for growth and maintenance. But we need to eat to fulfil all our requirements, and to have a reserve which we can call on when food supply is scarce. As mentioned above, areas of the brain, the liver and the stomach send signals when our store of essential nutrients is low. There are two 'warehouses' that hold these stores.

After a meal containing carbohydrates is eaten and digested, glucose levels rise, and the pancreas secretes insulin, which stimulates enzymes to convert glucose into glycogen. This is stored for future use in the short-term warehouse located in the cells of the liver and muscles. After about four hours from eating, or when we carry out strenuous exercise, the body needs replenishing, and the glycogen is synthesised and converted into glucose. The brain and the central nervous system receive glucose from the liver. The glycogen in muscles is used only for local use.

If we do not eat for long periods, the central nervous system utilises the long-term warehouse containing adipose fat tissue. This tissue is found under the skin and in the abdominal cavity. The cells convert nutrients and store them. The tissue can greatly expand causing overt obesity in some people. Adipose fats contain glycerol and fatty acids, and these are used by the rest of the body in times of need.

There are several reasons why we start to feel hungry, some of which are not necessarily related to the level of nutrients in our body. The sight or smell of food can produce feelings of hunger, as can looking at people eating around a table in a restaurant. Our culture tends to have fixed times for meals throughout the day, and our bodies become accustomed to being replenished at these periods. Therefore social reasons often outweigh metabolic necessity. If left to our own devices, we would probably eat smaller amounts at irregular intervals as instructed by our brain, and according to need.

The feeling of hunger is triggered by a low level of glucose, which is the primary fuel to the brain. The brain sends signals to the pancreas to secrete insulin and converts glucose for storage in the fatty tissues. This depletes the levels of glucose available in the bloodstream for use by the brain. Therefore a series of reactions are set in motion to cause us to eat. For example, as mentioned before, the hormone ghrelin secreted by the stomach and upper intestine stimulates appetite when nutrient levels are low.

What causes the stomach to rumble?

Borborygmi is the technical name given to the phenomenon. It comes from a Greek word meaning rumble. We swallow gases from the air and in food and drink, and some are also produced during the digestive process. As the gases travel through the liquids in the system, they produce the growling sounds. But it is only when the stomach contracts as it empties, that the sounds become amplified, and can be heard.

What makes us feel thirsty?

68% of body weight is water, which is received from drinking and the water content in food[2]. We may not realise the total water content in some foods. For example over 70% of the content of a potato is water. Water is lost through perspiration, when we breathe out and in urine.

The evolutionary process has adapted our cells to living on land from their original sea-based environment. There are regulatory mechanisms in place to maintain a balance between the sodium and water contents in our cells. Most of the water in the body is contained in fluid inside cells (intracellular). The rest is shared among fluid between cells (interstitial) and blood plasma.

There are two different conditions that make us feel thirsty. The first is a loss of blood or fluids caused by vomiting or diarrhoea. This reduces the blood flow to the kidneys, and secretes an enzyme called renin that reacts with other chemicals to stimulate the pituitary gland at the base of the brain. Renin secretes a hormone called vasopressin that constricts the blood vessels and instructs the kidneys to reduce the flow of water to the bladder. This makes us feel thirsty, encouraging us to start drinking.

The second reason is from eating a salty meal, which increases the level of sodium in the body, and upsets the normal balance between sodium ions and water between the cells. Special receptors in the brain called osmoreceptors respond to changes in the concentration of the solution, and osmotic pressure pulls water out of the cells in an attempt to restore the balance. When this happens, information is transmitted to the brain to encourage us to drink.

Why do we like or dislike certain foods?

Humans generally prefer sweet tastes. These indicate the presence of sugars which are used by the body to produce energy. Fruits and berries have the most nutritional value when ripe, and this is also when they taste the sweetest. Saltiness shows that sodium chloride is in the food. Reaction to these tastes depends on the state of the body. If we release a lot of sodium through sweating, then salt receptors become less sensitive and we can tolerate more salt in our food to make up the deficiency.

After a large meal, food loses its appeal, reinforcing the signals from the brain that the body has enough nutrients to meet its needs.

Humans learn from experience or teaching that bad or poisonous foods taste bitter or sour. This is because bacterial activity produces acidity, so we tend to avoid these tastes whenever possible. Food that smells bad, such as rotten eggs or meat is avoided. Some plants produce poisonous alkaloids which taste bitter. This protects the plant from being eaten. The large flower called stinking corpse lily smells like decaying flesh. As well as deterring predators, its odour and appearance also encourage carrion-eating flies to land on it, and inadvertently pollinate its flowers.

Memory plays a role in our food preferences. Studies have shown that people develop an aversion to certain foods when it is associated with a bad experience. People undergoing chemotherapy often show distaste for foods they have eaten just prior to the therapy, even though there is no biological connection. This is because the drugs induce nausea, which then forms a subconscious association with the food. Eating bad food also produces nausea.

Social factors also affect our preferences. Children will dislike foods that other children dislike. As with other consumer items, if a food is in short supply, the demand will increase. Also most people prefer a varied diet and like to experience new tastes. Biologically, this has the advantage of providing most of the vitamins and minerals we require.

There are about 10,000 taste buds in the mouth, most of which are arranged around the tongue. The tip of the tongue is most sensitive to sweet and salty tastes, the sides to sour and the back to bitterness. Taste receptors form connections with neurons and transmit information to the brain.

Some people have a larger than normal number of taste buds in a certain area of the tongue, which makes them especially sensitive and intolerant to the taste (see chapter on Taste and Smell).

Why do we need to smell food?

The sense of smell helps to identify food. The smell receptors have cells which lie in the mucous lining at the base of the brain. There are over 6 million receptor neurons in an area of 2 square centimetres of this lining. Other mammals such as rabbits make more use of their sense of smell, and have about 40 million receptor neurons.[3] We need to sniff to pull the air upward into the nasal cavity to reach the receptors. The life cycle of these cells is about 60 days and new ones are continually being formed. Some of the nerve endings are sensitive to pain, as when for example we sniff an irritant such as ammonia. Most of the foods we smell are made up of a variety of chemicals. However, we do not differentiate between these, identifying the product alone such as coffee or cured bacon at breakfast.

Our sense of smell appears to have degenerated throughout history, as it becomes less important to us than our sense of vision to react to food. Out of approximately 900 olfactory receptor genes, only 350 appear to be functional[4].

What makes us stop eating and drinking?

The brain sends us various signals along the digestive system that tells us when we've had enough.

Receptors in the eyes, nose, tongue and throat tell us about the food and drink we are consuming, but this needs to be learned. If we overeat then we become nauseous. We learn the approximate amounts that are right for our bodies, although a very varied diet can make us regularly eat more than necessary, leading to obesity.

When we have finished a meal, food in the intestine triggers the release of the hormone CCK, which notifies the brain that there is enough food in the body. CCK sensitises stretch receptors in the stomach, producing the feeling of fullness. It also causes the muscle between the stomach and the duodenum to contract, preventing more food from entering from the stomach, effectively slowing down the digestive process. Studies with rats[5] confirm that food intake can be adjusted to compensate for the loss or gain of calories, to maintain a regular intake.

Other experiments suggest that the liver contains detectors that are sensitive to glucose, and sends signals to the brain thereby inhibiting further eating.[6] This follows signals received by the liver from the stomach and the duodenum.

Finally the peptide hormone YY produced in the intestines inhibits appetite in response to the intake of food.

Why do some people under- or over-eat?

There may be many reasons why people eat abnormally, but stress and anxiety and low self-esteem seem to be common factors that cause us to ingest abnormal amounts of food.

There are two main conditions associated with the phenomenon that causes weight loss: anorexia nervosa and bulimia nervosa. Anorexics have a distorted negative body image, and perceive themselves to be much larger than they really are. The group most vulnerable to the illness are young women from teenage to 30s. Some studies suggest that media pressure connecting thinness to attractiveness may be an initiating cause[7].

Other studies suggest that anorexics share a genetic risk with clinical depression.[8] The neurotransmitter serotonin has also been implicated as a possible cause of the problem. A receptor linked to the regulation of feeding, mood and anxiety produced low levels of serotonin. Studies indicate that the serotonin system is impaired even after recovery from anorexia, indicating an underlying risk if untreated.[9] (see chapter on Hormones).

Anorexics often have personality traits that predispose them to the condition. These include an obsessive need for strive for perfection, an inflexible attitude and need to be in full control.[10]

Bulimics take a different approach to remaining thin, but the outcome is similar. They will eat large quantities of food and then vomit or take laxatives to prevent gaining weight. Both of these conditions can be fatal.

When food supplies are scarce, the body obtains its fuel from internal fat stores. But where food is in abundance, some people over-eat because of psychological or cultural pressures, and this leads to obesity. In Western societies, it is more convenient for busy people to eat meals at regular times throughout the day regardless of need, and the body will continue to store fat as a result.

There may also be genetic factors, as obese parents tend to have obese children. Parents also encourage their children to eat everything on their plate. Meals arranged in several courses also encourages over-eating.

It is generally believed that people with a poor metabolism gain weight. This is untrue. On the contrary, it is people with an efficient metabolism who readily accumulate fat. Those with inefficient metabolic rates can eat large quantities without putting on weight. Therefore it is an advantage to have an slow, inefficient metabolism in places where food is abundant.

Binge eating appears to reduce anxiety, and people with the disorder seem unable to recognise the signals of hunger and satiety. The Department of Health in England has forecast obesity levels will have risen to 12 million adults and 1 million children by 2010.[11]

Which nutrients does the body need to thrive?

The body needs carbohydrates, proteins, fats, vitamins, minerals, water and salt to thrive.

It is recommended that up to 60% of our total daily calorie intake should consist of carbohydrates. Many of these foods are rich in starch and fibre, which aid the digestive process, and are found in bread, potatoes, rice, fruits and vegetables.

Another name for carbohydrates is saccharides from the Greek word for sugar. Table sugar is converted into glucose and fructose by an enzyme in the small intestine and absorbed into the bloodstream.

Proteins are found in meat, eggs and beans. The large molecules are first broken down by enzymes in the stomach. In the small intestine several enzymes break down the protein molecules even further into amino acids. These are carried in the bloodstream to all parts of the body and help to maintain cells.

THE DIGESTIVE SYSTEM

Fat molecules, as previously mentioned are a rich source of energy for the body. The liver produces bile acids which dissolve the fat in water, and enzymes break down the fat molecules into fatty acids and cholesterol. Much of it is stored for later use.

Minerals are inorganic substances which cannot easily be broken down. They come in two groups. We need about 100 milligrams of major minerals per day. The first group include calcium, phosphorus and potassium, which are found in milk. Magnesium is found in fruits, vegetables and cereals, and sulphur is found in meat.

The second group is trace minerals and we need even smaller amounts of these. They speed up chemical reactions in the body. The main trace minerals are iron, iodine, zinc, fluoride, selenium, copper, chromium, and manganese, and these are found mainly in bread, cereals and grains, and also in meats and fruits.

Minerals such as calcium form part of the structure of bones and teeth. They also help release energy from food. Iron is used to form the haemoglobin part of the red blood cells, carrying oxygen around the body. But too much iron can accumulate in the liver and cause a toxic reaction.

Vitamins are organic substances containing carbon. They also divide into two main groups. Fat-soluble vitamins are stored in fat tissues in the body, until it needs to use up some fat. They include vitamins A, D, E and K.

Vitamin A, also known as retinol, is essential for vision, growth and development. Too much can be toxic. It can be found mainly in sweet potatoes and carrots, also in spinach, apricots, eggs and milk.

Vitamin D absorbs calcium mainly from sunlight. It regulates bone formation and can also be found in dairy foods and fatty fish.

Oxidants are chemicals that transfer oxygen atoms. Along with vitamins A and C, vitamin E is an antioxidant involved in cell metabolism. Antioxidants reduce cell damage by preventing oxidants from causing harm. Antioxidant-rich diets are thought to slow down the ageing process. Whole grains such as wheat and oats, leafy green vegetables, sardines and nuts all contain vitamin E.

Vitamin K controls blood clotting and bone development and repair. It can be found in leafy green vegetables, cheese, yoghurt and green tea.

Water-soluble vitamins are used up as they travel in the bloodstream. Any unused vitamins end up in the bladder and are flushed away. Vitamin C and the B group of vitamins are included in this group. Also in this group is folic acid and biotin.

Folic acid helps to maintain a healthy heart and is recommended in the early stages of pregnancy, as it assists cell growth and reduces the risk of birth defects. It is found mainly in green vegetables such as broccoli.

Biotin improves metabolic reactions and synthesises fatty acids. Lack of biotin can lead to hair loss and skin infections. Important sources are yeast, calves liver, rice and peanut butter.

Vitamin C helps to form collagen, which protects bones and muscles. It also absorbs iron. Citrus fruits, strawberries, tomatoes, broccoli and cabbage are all rich in this vitamin.

The B vitamins include thiamine, riboflavin, niacin, pyridoxine and B12. They provide energy to the body by converting glucose from carbohydrates. They also metabolise fats and proteins and maintain a healthy nervous system. B vitamins can be found in wheat and oats, fish, poultry and meats, dairy products, green vegetables, and citrus fruits.

As discussed earlier, water and salt in the correct proportions are essential elements for the maintenance of bodily functions.

Why doesn't dieting work for some people?

People adjust their metabolic rate according to the amount of calories being ingested. Therefore, when a person begins dieting, the metabolic rate falls preventing weight loss.

A study was carried out with obese people who normally have an intake of 3500 calories per day[12]. They were restricted to a diet of 450 calories per day, which was a drop of 87%. However the basic metabolism also reduced by 15%. After three weeks on the diet, average body weight had only reduced by 6%. Another study has shown that when the calories of a normal lunch are changed, people will unconsciously adapt their intake for the rest of the day to restore their total daily intake to normal levels[13].

Dieting can be counterproductive. People who regularly diet and then relapse can change their metabolism. Eating very little can improve the efficiency of their metabolism, which means that they will quickly store fat. An experiment was carried out on a group of wrestlers who needed to fast quickly to qualify them to be included in a lower-weight group[14]. They then binged to regain weight fast. It was noted that their resting metabolic rate was 14% lower than another group which took part in the experiment but acted as a control by not altering their eating behaviour.

Disorders affecting the digestive system

Acid Reflux, also known as GORD (gastro-oesophageal reflux disease) occurs when the some of the contents of the stomach mixed with the gastric juice hydrochloric acid are regurgitated into the oesophagus. The acid causes inflammation.

Appendicitis is caused when the appendix becomes enlarged and inflamed, causing pain. In its chronic condition it is sometimes known as a Grumbling Appendix. In its acute form it must be removed by surgery.

Cirrhosis of the liver is usually caused from excess alcohol consumption over a long period. The cells of the liver become damaged, causing internal scarring and harden. It is no longer able to remove toxic chemicals from the blood and is progressive. Liver transplants may be undertaken.

Coeliac Disease is caused when the lining of the small intestine is damaged, due to hypersensitivity to the gluten protein in wheat, rye and other cereals. The patient loses weight and can become anaemic. The condition tends to run in families.

Colitis is an inflammation of the colon and causes diarrhoea with blood and mucus. It can be caused by a bacterial or viral infection.

Crohn's Disease is an inflammation of the small intestine in younger people, but it also affects the rectum in the elderly.

Diarrhoea is increased fluidity and elimination of faeces. It causes dehydration and weakness from the excess loss of fluids and nutrients. It may be triggered by an infection or eating bad food.

Diverticulitis is an inflammation of the diverticula in the colon. These are small pouches of mucous. It can cause intestinal bleeding or a fistula. It is thought the cause may be connected with a deficiency of dietary fibre, or inefficient peristalsic action.

Flatulence is caused by air in the stomach or intestines which has been introduced when swallowing food or drink. Some foods, such as pulses produce gas during the digestive process.

Gallstones are masses of cholesterol and bile pigments which accumulate in the gallbladder or in the bile ducts. A diet high in fat may be a contributing factor. They cause problems when they become too large to pass through the gallbladder.

Hernia are protrusions of an organ through weak muscle tissue. A hiatus hernia is caused when the stomach protrudes through the diaphragm into the chest. Abdominal hernia can be caused from lifting heavy objects, or straining when constipated.

Hiccups are repeated, involuntary contractions of the

diaphragm. These are followed by the vocal cords closing quickly, causing the 'hic'.

Irritable Bowel Syndrome (IBS) causes bouts of diarrhoea and constipation and pain. It is more common in women. Anxiety and stress appear to exacerbate the symptoms.

Jaundice produces a yellow discolouration of the skin and the whites of the eyes. It is caused by an accumulation of bilirubin in the blood. Bilirubin is formed from haemoglobin during the breaking down of old red blood cells in the spleen. It is absorbed by the liver and excreted in bile and urine. Babies may be born with jaundice caused by the immaturity of the liver. The condition usually clears up within a week and is harmless.

Piles, also known as haemorrhoids are swollen veins in the lining of the anus, which can bleed and protrude outside the anal canal. Straining when constipated is a likely cause, and pregnant women are vulnerable to the condition.

Thrush is an acute fungal infection of the mouth caused by yeast. It usually affects people with suppressed immune systems through antibiotics or steroids. It can also occur in bottle-fed babies.

Ulcers are protrusions on the surface of parts of the digestive system such as the mouth or stomach. They are caused by an overproduction of acid in the gastric and intestinal juices. Ulcers can also occur on the skin.

Disorders affecting the urinary system

Cystitis is an inflammation of the lining of the bladder usually caused by a bacterial infection. The symptoms are a frequent urge to urinate, which is accompanied by a burning sensation. It is more common in women because the urethra is shorter and the bacteria can more easily reach the bladder.

Kidney failure occurs when the kidneys become diseased. The body can function with one kidney only, but if both need to be removed, a dialysis machine is necessary to filter the blood in place of the failed kidneys.

Kidney stones (calculus) can form from calcium oxalate and other crystallised salts in the urine. They cause an obstruction in the urine flow. Small stones usually pass out of the body in the urine and large stones can be surgically removed.

Nephritis is an inflammation of the glomerulus inside the kidney. It may be caused by an infection, or an abnormal immune response. Gout can also produce nephritis.

REFERENCES

[1] Waugh, A., Grant, A., (22006), *Ross and Wilson Anatomy and Physiology in Health and Illness*, Churchill Livingstone Elsevier, p.339

[2] Toates, F., (2001), *Biological Psychology*, Pearson Education Limited, p.430

[3] Rosenzweig, M.R., Breedlove, S.M., Watson, N.V., (2005), *Biological Psychology*, Fourth Edition, Sinauer Associates, Inc., p. 273

[4] Glusman, G., Yanai, I., Rubin, I., Lancet, D., (2001), The complete human olfactory subgenome. *Genome Research*, **11**, pp. 685 – 702

[5] Mather, P., Nicolaïdis, S., Booth, D.A., (1978), Compensatory and conditional feeding responses to scheduled glucose infusions in the rat. *Nature*, **273**, pp. 461-463

[6] Tordoff, M.G., and Friedman, M.I., (1988), Hepatic control of feeding: Effect of Glucose, fructose and mannitol. *American Journal of Physiology*, **254**, R969-R976

[7] Tiggerman, M/., and Pickering, A.S., (1996), Role of television I adolescent women's body dissatisfaction and drive for thinness. *Internationat Journal of Eating Disorders*, **Sept.20**, pp. 199-203

[8] Wade, T.D., Bulik, C.M., Neale, M., Kendler, K.S., (2000), Anorexia nervosa and major depression: shared genetic and environmental risk factors. *American Journal of Psychiatry*, **157**, pp. 469-71

[9] Kaye, W.H., Bailer U.F., Frank, G.K., Wagner, A., Henry, S.E, (2005) Brain imaging of serotonin after recovery from anorexia and bulimia nervosa, *Physiology & Behaviour*, **86**, pp. 15-17

[10] Wonderlich,S.A., Lilenfeld, L.R., Riso, L.P., Engel, S., Mitchell, J.E., (2005) Personality and anorexia nervosa. *International Journal of Eating Disorders*, **37** *Suppl*. P.68-71

[11] Department of Health Forecasting Obesity by 2010, 25 August 2006

[12] Bray, G.A., (1969), Effect of calorific restriction on energy expenditure in obese patients. *The Lancet*, **2**, pp. 397-398

[13] Foltin, R.W., Fischman, M.W., Moran, T.H., Rolls, B.J., and Kelly, T.H., (1990), Caloric compensation for lunches varying in fat and carbohydrate content by humans in a residential laboratory. *American Journal of Clinical Nutrition,* **52**, pp. 969-980

[14] Steen, S.N., Oppliger, R.A., Brownell, K.D., (1988), Metabolic effects of repeated weight loss and regain in adolescent wrestlers. *Journal of the American Medial Association,* **260,** pp. 47-50

CHAPTER 5

DRUGS and ADDICTION

How do drugs act on the brain and body?
Why do some people become addicted to drugs?
What are the effects of some common psychoactive drugs?

How do drugs act on the brain and body?

Drugs that affect the brain are called psychoactive. They interact directly on the central nervous system. Drugs alter the transmission of neural signals to the brain by changing the rate of release of neurotransmitters at the synapses (junctions) between different nerves. The effects are produced in two ways: by agonist drugs that allow more neurotransmitter to be released, or by antagonist drugs which block the receptor cells to inhibit its release.

Drug molecules that interact with receptors are called ligands. As agonists, they start off the process of release, as antagonists they prevent the process, which has effects on other neurotransmitters. A molecule of a drug will usually travel around the body until it meets a specific kind of receptor, like a key fitting a particular lock. There are some drugs that fit several kinds of receptor, and these can cause various types of reaction, depending on the strength of the drug, and the natural function of the neurotransmitter. Table 5.1 summarises the effects of some drugs on transmission.

A neurotransmitter interacts with receptors in different ways. One method is to open an ion channel to allow the signal to move along towards its target cells. These receptors are called ionotropic, and the signal fires quickly. Another type of receptor changes the structure of the target cells. These are called metabotropic and act rather more slowly.

Acetylcholine (ACh) is a transmitter that has pathways throughout the nervous system. There are two main types of ACh receptor called nicotinic and muscarinic.

Table 5.1 Effect of drug on neurotransmitter

DRUG	EFFECT ON TRANSMITTER
Agonists	
Cocaine	Blocks reuptake of dopamine
Black widow spider venom	Stimulates release of Acetylcholine (ACh)
Antagonists:	
Botulinum toxin	Inhibits release of ACh
Curare toxin	Blocks ACh
Reserpine	Creates monoamine vesicle leak

Nicotinic receptors are ionotropic and are found mainly on muscle end plates. They are stimulated by nicotine and ACh. The drug curare blocks the receptors. Jungle tribes in South America used the drug to coat the tips of arrows and darts to paralyse prey.[1] They are also found on ganglia in the autonomic as well as the central nervous systems, and can be blocked by the drug hexamethonium.

Muscarinic receptors are metabotropic and are stimulated by the drug muscarine that is found in some types of mushrooms. They mimic the action of ACh, and are established throughout the nervous system. They are blocked by the drug atropine, which is extracted from the plant atropa belladonna, also known as deadly nightshade. Nicotine and muscarine are both very toxic in their natural forms.

ACh receptor agonists are used to treat the disorder myasthenia gravis, which produces muscle weakness and destroys muscle end plates. They are also used to treat Alzheimer's disease by reducing the tangles of neurofibres in the brain that occur with this condition.

Other main classes of neurotransmitter include catecholamines such as dopamine and noradrenaline, and indoelamines such as serotonin and melatonin. Dopaminergic neurons are found in most regions of the brain, and affect many physiological procedures. Morphine and alcohol both increase the rate of dopaminergic neuronal activity. Serotonergic neurons are more thinly spread throughout the brain, and affect mood and sleep patterns. Glutamate is an excitatory neurotransmitter that is produced in excess when the brain receives a trauma, increasing the effects of the injury. Gamma-aminobutyric acid, known as GABA, has three groups of receptor that inhibit synaptic reaction both ionotropically and metabotropically.

There are four main groups of psychoactive drugs: depressants, stimulants, opiates and hallucinogens. All affect various regions of the brain to alter perception and behaviour.

Depressant drugs slow down the activity of the central nervous system. They increase the activity of GABA which calms and relaxes and help to relieve the symptoms of insomnia and stress. The most common depressant substances are ethanol, also known as alcohol, barbiturates and tranquilisers. Excessive, prolonged use can lead to liver and brain damage, amnesia and problems with the digestive system.

Stimulants have the opposite effect. They increase activity of the sympathetic and the central nervous systems, to produce greater attention and responsiveness. The most common stimulant drugs are caffeine, amphetamines, nicotine and cocaine. Except for caffeine which has far less severe reactions, excessive use of stimulants can lead to chronic insomnia, psychosis and paranoia.

Opiates slow down the rate of neural transmission in the central nervous system to suppress pain, lower blood pressure and respiratory rates. They also affect the activity of the intestines. Common opiates are opium itself, morphine, heroin, methadone and codeine. Taken in excess over long periods these drugs can lead to a depressed sexual drive and general feelings of tiredness and poor health.

Hallucinogenic drugs distort our perception of reality, and can lead to aggressive behaviour and convulsions. Common drugs are LSD, mescaline and cannabis (also known as marijuana, hashish or ganja). Continued, excessive use can lead to psychosis and long-term disruption of mental processes.

Why do some people become addicted to drugs?

There are some types of personality that are prone to addiction. This can be to gambling, shopping or the internet as well as to different types of drug, including alcohol. The reasons for becoming addicted also vary, but inevitably they are a form of escape from reality, or what is perceived to be reality. Social deprivation and depression, as well as the availability of drugs in the local environment also help to develop addiction. Low self-esteem coupled with an inability to feel part of normal society often leads to various levels of drug abuse. As the nervous system adapts to receiving the drugs, it becomes desensitised, and often needs increasing doses to produce the desired 'feel-good' impact. Some drugs alter the neural pathways leading to the brain. Cocaine affects the release of dopamine in this way.[2] The liver can break-down drugs more efficiently which means they are not as effective when they reach the brain. Therefore, increasing doses are needed to produce the same pleasurable feelings. All these consequences can lead from abuse to addiction.

Some people begin taking drugs to become accepted within their peer group. They are particularly sensitive to social signals in their environment. Smoking cigarettes or drinking alcohol is often started because friends and family are regular users.

There have been various studies with adopted children and identical twins indicating that the addictive personality is inherited, and this is separate from environmental influences.[3] In another study, pairs of identical male twins were compared with pairs of fraternal twins. They were asked to state their preference or otherwise for smoking marijuana. The study concluded that the identical twins gave more similar answers, and this indicated a genetic factor. This is because identical twins share exactly the same genes, and fraternal twins have only 50% identical genes.[4]

Another study compared the DNA from abusers and non-abusers, and found a gene that produces an enzyme that breaks down dopamine, and other similar substances. In its highly active form, it is found more often in drug abusers.[5] However, there is still no definitive link and research continues. If such a link is established, it will pave the way for therapies to combat addiction.

What are the effects of some common psychoactive drugs?

Stimulants: cocaine, caffeine, amphetamines, nicotine.

Cocaine is derived from the coca plant. The drug blocks receptors in the gum, preventing pain signals reaching the brain. Cocaine acts as an indirect agonist, causing an increase in dopamine levels at the synapses. The forebrain that receives dopaminergic neurons and areas of the forebrain that are affected, are the nucleus accumbens, the prefrontal cortex and the thalamus. The forebrain controls cognitive functions including perception, memory, reasoning ability, emotions and abstract thought as well as activities connected with movement, and patterns of eating and sleeping. Many of these areas of behaviour are therefore affected by cocaine use. Recent studies have also implicated the disruption of the neurotransmitter serotonin caused by the drug.[6] Cocaine has been used as a local anaesthetic by dentists in various forms such as novocaine or lidocaine. The drug blocks receptors in the gum, preventing pain signals reaching the brain.

Caffeine is found in many plants, but humans take it mainly as infusions from the beans of the coffee plant, and from the leaves of tea bushes. It is quickly absorbed into the body and the effects can be felt within an hour. It stops us feeling sleepy and helps in concentration and attention. The drug acts as an antagonist of the neurotransmitter adenosine, inhibiting its production. This results in increased dopamine activity which stimulates the brain. There have been many studies to show the effectiveness of the drug. One study of trained runners showed a 51% increase in cycling endurance after a dose of 9 milligrams of caffeine per kilogram of body weight.[7]

There are withdrawal effects when caffeine use is stopped. Adenosine helps to regulate blood pressure by dilating blood vessels. If the body is used to a certain level of caffeine, withdrawing it will increase the levels of adenosine, causing more blood to enter the brain through dilated blood vessels. This can cause headaches and nausea. Serotonin levels are increased with the use of caffeine, and the reduced levels after withdrawal of the drug can result in depressed changes of mood and lack of motivation. These effects are comparatively short-lived until the body reverts to natural levels of neurotransmission.

Amphetamines have a similar effect to cocaine as they increase levels of dopamine, as well as noradrenaline, and other associated transmitters. They increase alertness and reduce appetite. For these reasons they have been used medically in the treatment of attention-deficit hyperactivity disorder (ADHD) in children, as it helps them to concentrate and pay attention. The drug has also been widely subscribed as an aid to losing weight. However, the cocktail of weight loss and 'feel-good' factor became addictive and dangerous, and amphetamine use is now strictly controlled in most countries. Long-term use can lead to insomnia, depression and reduced mental ability, as well as causing damage to the heart, liver and lungs.

Nicotine is extracted mainly from the leaves of the tobacco plant. It is also found in very small quantities in other plants in the nightshade family such as aubergine, tomato and green pepper. When taken to excess the drug can result in cancer of the lungs, mouth, throat and oesophagus. The drug binds to nicotinic ACh receptors and enters the blood stream from the comparatively large lung surface, where over time it leaves a residue of tar. It indirectly stimulates the adrenal glands which release epinephrine (adrenaline). This increases heart rate, and the feelings of arousal. Nicotine can reach the brain within 10 seconds, so its stimulatory effects are virtually immediate.

Nicotine is very addictive, which is why cigarette smoking is one of the two most widely used drugs in the world (the other is alcohol). The neural circuits in the brain are altered to receive higher levels of dopamine. As a stimulant, it increases blood pressure and constricts blood vessels. Cholesterol and fat enter the blood stream and clog the arteries, placing great strain on the heart.

Sedatives: ethanol (alcohol), barbiturates, tranquilisers.

Alcohol taken in moderation produces a feeling of well being and lowers anxiety levels. Studies indicate that it reduces the risk of heart attack[8], raises protective (so called 'good' high density lipoproteins) cholesterol levels and lowers blood pressure.

Alcohol activates GABA-benzodiazepine receptors, making them more sensitive. It indirectly increases the release of dopamine in the nucleus accumbens area of the brain. Alcohol also binds to and inhibits glutamate receptors in the brain. However, in high doses and taken over the long term, alcohol is very detrimental to health and can be fatal. Its abuse leads to aggressive, antisocial behaviour. The changes in brain function can lead to shrinkage of the brain itself, and initiate insomnia, headaches and memory lapses. Prolonged use damages the liver, cardiovascular and gastrointestinal systems. Alcoholics are prone to lung and muscle deterioration and skin disorders. Reduced sperm count and shrinkage of the ovaries and testicles can bring about infertility. Women who drink too much during pregnancy can cause permanent damage to the foetus producing stunted growth and reduced mental development. Alcohol has been the drug of choice over many generations and civilisations throughout the world.

Barbiturates bind to ionotropic GABA receptors that are ligand-gated chloride channels ($GABA_A$ group). They depress the activity of the reticular formation, a set of nuclei in the brainstem that is involved in sleep and wake patterns. They are often used as a medication to promote sleep. Taken in excess, they can stop the normal functioning of the medulla oblongata. This organ is in the lower part of the brainstem and relays information between brain and spinal cord. It controls autonomic functions including breathing, heart rate, blood pressure and swallowing. If taken in conjunction with alcohol barbiturates cause death by suffocation or heart failure.

Tranquilisers, whilst having similar actions to barbiturates, differ in that they do not induce sleep. They mainly belong to a group of substances called benzodiazepines, which include the prescribed drugs Valium® and Librium®. They also act on $GABA_A$ receptors, enhancing their inhibitory effects. This leads to a decrease in firing rate of signals to the brain. The result is to produce feelings of relaxation. The drugs are often prescribed to treat symptoms of paranoia and hallucinations.

Opiates: opium, morphine, heroin, methadone, codeine.
Opiates, also known as opioids, are substances that have been extracted from the resin of seed pods of the opium poppy plant. They are very effective pain killers, as they depress the nerve signals to the brain. Also used illegally in some forms to achieve feelings of euphoria. The nervous system quickly becomes tolerant of the various forms of the drug, which is why it is so addictive, and the user needs increasing doses to achieve the same results. Opiates stimulate special opioid receptors, of which there are three main types, μ (Greek letter mu), κ (Greek letter kappa) and ∂ (Greek letter delta). They react differently depending whether the opiate is an agonist or antagonist. They are all G-protein-coupled metabolic receptors, which are embedded in the plasma membrane of neurons. G-proteins bind to guanine nucleotides. When the opioid binds to the receptor, the G-protein is activated and it diffuses across the membrane, inhibiting neurotransmission. GABAergic neurons are depressed, which allows dopaminergic neurons to produce more signals to the brain. This releases more dopamine in the nucleus accumbens producing the 'highs' associated with using opiates.

The body also produces its own endogenous opioids such as endorphins, enkaphalins and dynorphins that act through different opioid receptors. They affect our reactions to pain and also regulate other systems such as mood, hunger and thirst. The use of exogenous (outside the body) drugs, such as opium and its derivatives serve to enhance the effects of these naturally occurring substances.

Opium is a narcotic – that is it induces feelings of numbness.

Morphine is the principal agent of opium and relieves pain. It reacts with the µ-opiod receptors that are scattered throughout the brain, but especially in the amygdala, hypothalamus and thalamus. It particularly affects the nucleus accumbens region. The drug also interacts with κ-opioid receptors found in the spinal cord.

Heroin is chemically processed from the opium poppy, and is also used as a pain killer and recreational drug. It is synthesised from morphine. Methadone is synthetically produced, and is used medically as an analgesic. It helps to wean addicts away from heroin as its effects are similar, but not addictive. Codeine is also used medically to relieve pain. It can be extracted from opium or synthesised from morphine. Enzymes in the liver convert it to morphine. Its effects are milder and therefore it is less addictive than morphine

Hallucinogens: LSD, ecstasy, mescaline, marijuana.
Hallucinogens distort sensory perceptions to produce psychedelic effects and delusions. They warp the perception of the surrounding environment.

Lycergic acid diethylamide (LSD), also known colloquially as acid, is found on grains in the fungus ergot. The drug binds to and activates serotonergic receptors to create the hallucinogenic effects in the brain.

Methylenedioxy amphetamine, also known as ecstasy activates serotonin receptors, which causes excessive amounts of dopamine to be released from dopaminergic neurons.

Mescaline is a natural substance mainly derived from some types of cactus. The drug affects G-protein-coupled receptors to change cellular structure and ultimately alter the normal firing rates of neural signals to the brain.

Marijuana is a tetrahydro-cannabinol drug (THC) derived from the cannabis plant. It binds to cannabinoid, G-protein-coupled receptors. However, the body has only a moderate tolerance to the drug, which is eliminated very slowly from the body. Because of this, small doses are usually sufficient to produce the desired effect.

The body also produces its own natural substances through cannabinoid receptors that are found in the cerebral cortex, substantia nigra and hippocampus areas of the brain. The hippocampus is the region of the brain that controls learning, which is why abuse of these types of drug often adversely affects short-term memory.

REFERENCES

[1] Carlson, N. R., (1994), The Physiology of Behaviour, Fifth Edition, Allyn and Bacon, p.72

[2] Ungles, M.A., Whistler, J.L., Malenka, R.C. and Bonci, A., (2001), Single cocaine exposure in vivo induces long-term potentiation in dopamine neurons. *Nature* **411,** pp. 583-586

[3] Cadoret, R.J., O'Gorman, T., Troughton, E., and Heywood, E., (1986), An adoption study of genetic and environmental factors in drug abuse, *Archives of General Psychiatry,* **43,** pp. 1131-1136
and
Goodwin, D., W., (1985) Alcoholism and heredity: A review and hypothesis. *Archives of General Psychiatry,* **36,** pp. 547-61

[4] Lyons, M.J., Toomey, R., Meyer, J.M., Green, A.I., Elsen, S.A., Goldbert, J., True, W.R., and Tsuang, M.T., (1997) How do genes influence marijuana use? The role of subjective effects. *Addiction* **92 (4),** pp. 409-417 In National Institute on Drug Abuse (NIDA) article 1997

[5] Vandenbergh, D., Rodriguez, L.R., Miller, I.T., Uhl, G.R., and Lachman, H.M., (1997) High-activity catechol-o-methyltransferase allele is more prevalent in polysubstance abusers. *American Journal of Medical Genetics,* **74,** pp. 439-442 in NIDA article, op.cit.

[6] Patkar, A.A., Berettini, W.H., Hoehe, M., Hill, K.P., Sterling, R.C., Gottheil, E., Weinstein, S.P., (2001), Serotonin transporter (5-HTT) gene polymorphis, and susceptibility to cocaine dependence among African-American individuals. *Addiction Biology* **6 (4),** pp. 337-345

[7] Graham, T.E., Spriet, L.L., (1991), Performance and metabolic responses to a high caffeine dose during prolonged exercise. *Journal of Applied Physiology* **71 (6)**, pp. 2292-2298

[8] Klatsky, A.L., (2004), Alcohol and Cardiovascular Health, *Integrative and Comparative Biology* **44**, pp. 324-328
and
Sacco, R.L., Elkind, M.,Boden-Albala, B., Lin, I-F., Kargman, D.E., Hauser, W.A., Shea, S., Paik, M.C., (1999), The Protective Effect of Moderate Alcohol Consumption on Ischemic Stroke, The *Journal of the American Medical Association,* **281**, pp. 53-60

CHAPTER 6

Emotions And Stress

What are emotions?

How many emotions do we have?

Why are facial expressions so important?

How do emotions develop?

Which areas of the brain are involved in emotional responses?

Why do we get angry?

Why do we cry?

Why do we laugh?

What makes us frightened?

Why do we blush when nervous or embarrassed?

What is stress?

Why does stress affect digestion?

How does Post-Traumatic Stress Disorder affect the body?

What are emotions?

Emotions are feelings that we experience in certain situations. These emotions can be positive, making us feel in a good mood, or negative, making us feel bad tempered. They can be separated into three characteristics: subjective, physiological and cognitive.

Our subjective emotions are private and can be seen through our facial expressions. We can also assess the emotions of others by studying their facial appearance. The physiological aspect is shown in the way our bodies react. For example, when we are frightened, the heart beats quicker, we perspire more and feel nauseous. We also tend to freeze when we hear a sudden loud noise, before deciding whether to stay or run if the situation becomes unsafe. The cognitive aspect is utilising our memory and understanding the situation in order to express our feelings.

There are several theories trying to explain how our emotions are formed. The James-Lange theory[1] suggests that a certain situation or stimulus will produce physiological changes, which result in the emotion. For example, if a person unexpectedly sees a friend across the road, the face breaks into a smile of recognition and the step may quicken. This causes feelings of happiness. In this case, the friend acts as the stimulus, and the brain interprets the actions of the muscles and reacts to produce the emotion.

The Cannon-Bard theory [2] challenged the James-Lange Theory[3] and suggests the emotion starts before autonomic changes, and that these physiological changes are similar in any strong emotion. The brain responds to the stimuli and sends signals to the rest of the body via the nervous system.

The Schachter and Singer[4] theory proposes that the brain interprets a situation and activates the body to react simultaneously. It recognises that the stimulus is not able to produce an emotion on its own – there needs to be the physical response as well.

Jaak Panksepp[5] advocates that mammals have genetically pre-wired circuits which interact with decision- making processes in the brain to organise a variety of behaviours. Sensory systems are changed to respond to emotionally charged situations which continue to exist long after the special circumstances that caused them.

These theories all accept that different emotions cause noticeable physical responses. For example, when we are frightened or angry, we feel the heartbeat racing and muscles tensing. When we are sad, there is a tight feeling in the throat and eyes, and our limbs become heavy. When we are happy, we feel lighter, move easier and the chest seems to swell.

How many emotions do we have?

It has been suggested[6] that emotions can be 'free-floating' or have an object in mind. Free-floating emotions include happiness, sadness, anger and fear. Those that always have an object in mind are love, disgust and contempt. Happiness occurs when we have achieved a goal, and a subjective plan of action would be to continue along the same path. Sadness is a result of failing to reach a goal or carry out a particular plan.

We need this emotion to actively seek to change this situation. Anger is evoked when we are frustrated in our goal, and we must actively do something to change the circumstances, which may include being physically aggressive. Fear is caused by a threat to our well-being or a conflicting goal.

Love is attachment to another person and requires actions such as nurturing, sexual activity or maintaining social contact. Disgust is caused by our senses coming into contact with a perceived contaminated object. Having contempt for another person means they can be treated without consideration.

According to Robert Plutchik[7] there are eight primary emotions which are grouped into four opposing pairs at three levels of intensity, high, medium and low. Table 6.1 shows the four pairs at each level. All other emotions arise from these pairs.

Table 6.1 Basic emotional pairs and intensity levels

HIGH LEVEL OF INTENSITY	MEDIUM LEVEL OF INTENSITY	LOW LEVEL OF INTENSITY
Ecstasy/Grief	Joy/Sadness	Happiness/Pensiveness
Adoration/Rage	Affection/Disgust	Regard/Boredom
Rage/Terror	Anger/Fear	Annoyance/Apprehension
Vigilance/Amazement	Expectation/Surprise	Alertness/Distraction

Plutchik believed these primary emotions have evolved as survival mechanisms.

Why are facial expressions so important?

Charles Darwin first suggested that human facial expressions are innate responses and also that they are universal[8]. Humans from very different cultures throughout the world can easily recognise emotions such as happiness or fear from the appearance of a face. Darwin also noted that children who were born blind and deaf still express their emotions using the same facial features as sighted and hearing people. Humans can recognise and empathise with others based on their interpretation of facial expressions, and also convey to others their own feelings. It is an important form of social interaction and bonding. Studies have shown that even very young babies will imitate the expressions made to them by adults.[9]

Facial muscles in humans receive their information from cranial nerve VII. This controls facial expression. Muscles pull on their attachments to the skin and can change the shape of the mouth, eyes and nose, lift eyebrows or wrinkle the forehead. The motor cortex which is responsible for all types of movement has a very large area responsible for the face. The cerebral cortex can direct movement of facial muscles on either side or both sides of the face, which is why we can voluntarily wink with one eye or move one side of the mouth or nose.

How do emotions develop?

Our emotional responses to new situations set standards for our future behaviour. They allow us to act instinctively and quickly when necessary. Evolutionary psychologists believe emotions are adapted through natural selection. Those that are successful, such as fear giving us the opportunity to choose between staying and fighting or running away as fast as we can, will be passed on to future generations. Our emotions support and reinforce rational thought.

Erik Erikson proposed his hypothesis, influenced by Sigmund Freud's psycho-sexual stages to maturity, that there are eight phases of social and emotional development throughout life.[10] Each one builds on the foundation of the previously learned phases. Table 6.2 outlines his theory.

Michael Lewis believes that most human emotions develop within the first three years of life.[11] At birth, babies are already primed with emotions of distress or contentment, interest or disinterest. By three months they can smile and be excited by familiar faces, or show distaste for some new foods. Between four and six months babies can show anger at being frustrated in a task. At six months they develop the emotion of surprise and by eight months they show fear at seeing strange faces or being unable to be with their familiar adults. Between eighteen and twenty-four months they exhibit self-awareness and can become jealous of siblings. By the age of three, they have begun to evaluate their behaviour against other children and display signs of pride, guilt or regret.

Table 6.2 Erikson's Stages of Emotional Development

PHASE	AGE	EMOTION
1	0-1	Learning how to trust or mistrust
2	1-3	Learning how to be autonomous – pride or shame at failure
3	3-6	Learning how to have initiative or guilt at failure
4	6-12	Learning to be industrious – self worth or inferiority at failing to achieve goals
5	12-18	Learning identity – self image or confusion and doubt
6	18-40	Learning intimacy – commitment or isolation at failure
7	40-65	Learning generativity – career, family, community or self-absorption
8	65+	Integrity – life's accomplishments or despair (dealing with loss and death)

Which areas of the brain are involved in emotional responses?

Emotion is the result of the interaction between the brain and the rest of the body. The autonomic nervous system produces effects connected with emotions, by altering the normal state of the body. For example, when we are afraid, sweat glands are stimulated to increase the supply of sweat, and bronchial tubes in the lungs expand to allow us to breathe faster. The system increases heart rate and directs blood to skeletal muscles by dilating large blood vessels. At the same time blood flow is reduced from other areas of the body by constricting the smaller blood vessels. All these processes prepare the body for flight or fight, when the brain gives the relevant signals – and all done in seconds! The endocrine system also produces hormonal changes throughout the body to be interpreted by the brain.

James Papez studied the post-mortem brains of animals and humans with emotional disorders and discovered that various pathways related to emotions.[12] They were subsequently called The Papez Circuit. Paul MacLean later suggested other areas be included, and that the whole should be called the Limbic System.[13] Figure 6.1 details these parts of the brain.

Figure 6.1 Limbic System

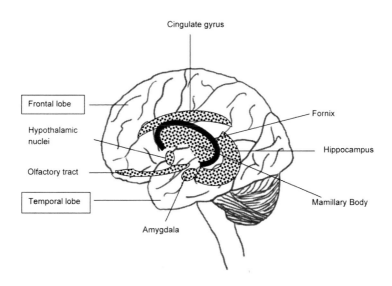

There is a difference of opinion between scientists regarding which parts of the brain should be included. Although the hippocampus was originally thought to be involved, it is now generally considered that it is not directly related. This is because injury does not appear to affect emotional responses, although it does involve memory which can affect some emotions.

The cingulate gyrus is part of the autonomic system and helps to regulate heart rate and blood pressure. It also processes information related to learning and memory.

The hypothalamic nuclei link the autonomic system to the endocrine system via hormone regulation.

The two mamillary bodies are important in forming memories and relaying signals between other areas of the limbic system

The olfactory tract is a pathway for the sense of smell.

The amygdala consists of groups of nuclei and various networks of pathways that send signals to and receive from different regions of the brain. These are especially related to the emotion of fear. It projects to the brainstem which affects behaviour, the hypothalamus which affects the autonomic system, and the cortex which relates to past experiences. For example, if bitten by a dog, the sound of barking alone can subsequently induce a feeling of fear. Studies have shown that people with damage to the amygdala appear to have a reduced capacity to recognise some emotions.[14] There are two separate pathways from the thalamus to the amygdala.

The fast route is direct between the two organs and works

unconsciously to tackle immediate threats. The slower route runs through the cerebral cortex, and from this we can make conscious decisions on a plan of action. Different pathways from various areas of the amygdala are stimulated depending on the type of motivation. For example, one pathway will activate the freeze response, another changes blood pressure and another will activate stress hormones.

The prefrontal cortex, the area at the very front of the brain, appears to play a vital role in regulating emotional behaviour. Damage to this part in humans changes the way they interact socially. An example of this is the famous case of Phineas Gage,[15] who, in the mid-1800s was involved in an accident at work. A dynamite explosion backfired, sending a steel rod through his right cheek which passed out at the top of his head, causing extensive damage to the frontal lobe of his brain. Remarkably, he survived and appeared physically none the worse after the accident. However, his personality had changed considerably. Before, he was hard-working, conscientious, polite and was well-liked by his colleagues. After the accident he became childishly irresponsible, obstinate and foul-mouthed. The damaged area meant he had less control over his emotions or the consequences of his actions.

Studying patients with brain damage and from brain scanning techniques has shown that the two cerebral hemispheres process information differently. The right hemisphere specialises in perceiving emotions in other people. It controls the left side of the face, which is more expressive than the right.[16] It is also more active in producing negative moods, whereas the left hemisphere is more active in producing positive moods.[17]

Why do we get angry?

Aggression can be physical or verbal. Aggressive behaviour is shown by most species as a survival mechanism, in order to obtain food, resist predators or get a mate. Human aggression is more complex, as it also encompasses the social standards of behaviour in different types of environment.

Certain conditions can trigger aggression. Alcohol and drugs weaken our ability to judge a situation accurately and less inhibited from taking action. Being in pain or discomfort can create an aggressive reaction. Frustration at not achieving a goal, or suspecting someone of thwarting your ambitions, can produce a violent reaction.

Males tend to show more aggressive tendencies than women, although it may be that women find less physical ways of expressing them. One reason that has been given for this is the level of testosterone in the body. High levels also increase sexual desire and correlates with male-to-male violence, in finding or keeping a mate. Women convicted of violent offences have been shown to have higher than normal levels of testosterone.[18] But other studies have shown that winning at competitive sports also increases levels of testosterone.[19] This does not lead to aggression but a greater desire to repeat the performance.

The neurotransmitter serotonin is involved in lowering levels of aggression. Higher levels of serotonin lower the tendency to behave violently.

It appears that aggression can also be learned and imitated.

There have been numerous studies indicating that children can be influenced to act aggressively after watching a violent film or playing an interactive violent computer game. A long-term study of over 700 families showed there were strong associations between the amount of time spent watching violent television as a teenager, and the likelihood of committing aggressive acts at a later date.[20]

Why do we cry?

There are three types of tears. Basal tears keep the eyes lubricated. The fluid is secreted by the lachrymal glands and contains water, glucose, urea, sodium, potassium, lipids and other chemicals which help to fight infection. The fluid produced rarely overflows. Reflex tears result from an irritation to the eye caused by a foreign particle such as a piece of grit, or substance such as the vapour that comes from peeling onions.

The third type is crying due to strong negative emotions such as stress, pain, fear or depression. We can also cry when we are very happy or when we laugh a lot. Under extreme stress, crying turns into sobbing which affects our breathing. The fluid overflows through the tear ducts at the corner of the eyes and also from the nose. The fluid in tears caused by strong emotions contains higher levels of proteins than the tears produced for lubrication, and it may be that tears release harmful, stress-related chemicals, which is why we cry.[21] The parasympathetic branch of the autonomic nervous system activates the lachrymal glands to produce tears.

However, another reason may be that crying is a form of

communication, inducing sympathy and reaction in others, which helps to reduce stress. For example, when children fall, they may not cry until noticing an adult is watching.

Why do we laugh?

There are many reasons why we laugh – and all of them seem good for our body and mind. There are three main categories: incongruity, superiority and relief. The incongruity theory is that we laugh at the unexpected result of a situational joke. We can predict the probable outcome, based on the story so far – but the end is completely different and we find that highly amusing. We often laugh at stupid or inappropriate actions. We feel superior and laugh at someone else's misfortune. The old slipping on a banana skin joke comes into this category.

The relief theory is that we laugh after a build up of tension at a situation. Many of us can relate to nervously laughing after being told of a serious incident or when attending an important interview. The laughter releases our inner stress.

Laughter is contagious – we often laugh when we see or hear other people laughing. This is why the producers of radio and television comedy programmes, in the absence of a live audience, will insert canned laughter – it is more likely to make the listener laugh too. When we laugh together, we feel happier and it helps us to bond with other people. We like to be with others with whom we can share this positive experience.

Gelotology is the name scientists have given to the study of laughter, and there have been many investigations into what happens to the body and which parts of the brain are involved. One study measured brain activity using an electroencephalograph (EEG) machine.[22] Unlike emotional responses which are generally connected to specific areas of the brain, it was established that the electrical patterns covered a circuit running through several different brain regions. It started on the left hemisphere which processed the words, and moved to the frontal lobe involved in emotional responses. There was then a spread of electrical activity between the left and right hemispheres, and on to the occipital lobe before finally stimulating motor areas to start the physical process of laughing – providing the signal was strong enough – in other words if the subject thought the joke was funny. All this was done in milliseconds.

Laughter can improve the immune system, by reducing levels of stress hormones which suppresses the system. Cortisol increases the response to stress, and studies have shown the levels in the bloodstream reduce after laughing.[23] Levels of salivary immunoglobulin A increase when we laugh. This helps to fight off infections breathed in through the mouth.

Laughter can also influence the endocrine system, which controls hormone activity. By changing secretion levels, all parts of the body are affected.

We also laugh when we are tickled. Our skin is sensitive to very light sudden touches, such as a little insect, which tends to cause us to brush it away as it could cause us damage. The theory is that if we know the cause is benign, we laugh in relief. So why can't we laugh when we tickle ourselves? Apparently the cerebellum in the brain distinguishes between movements that are expected or unexpected. When we laugh, we need an element of tension or surprise. Any action we take ourselves is expected and therefore cannot make us laugh.

What makes us frightened?

Fear is an evolutionary mechanism for survival. We become fearful when we are threatened by a dangerous situation. The degree of fear ranges from mild anxiety to extreme terror. Fear can be innate or learned. We have an innate fear of heights. Studies with babies have shown that they will not crawl over a part of the floor that looks as if it is a cliff edge. We can learn fear by experience or conditioning. For example, if there was an incident while learning to swim as a child, this could result in a fear of water. People often have unfounded fears of the unknown – such as alien beings from outer space, or other people from different cultures.

But our fears generally have a logical explanation, even though the level of fear may be irrational. When this happens, we develop phobias such as a fear of spiders. There are some areas of the world where some spiders are poisonous and there are sound reasons to be afraid. However, arachnophobes are scared of spiders they know are harmless. Phobias usually develop from childhood fears, and can arise when a child sees fear shown by other people. I have a dog phobia, and believe this stems from an incident that occurred when I was about 5 and walking with my mother. As we passed a dog, it started barking very loudly causing us to be very startled. I noticed my mother held my hand more tightly and looked frightened and this memory has always stayed with me. The unreasonable fear of flying may stem from other related fears of being in enclosed spaces or heights. But it may also be due to sensational media attention given to the comparatively rare plane crashes. Statistically, there are many more accidents caused by cars or even bicycles – but we rarely develop phobias about these means of transport.

As with other emotional states, it is the sympathetic nervous system and organs of the limbic system which control our reactions to fear. We inhale sharp intakes of breath, to provide more oxygen for the body, which is pumped to the muscles in readiness for attack or escape. Sweat glands secrete more fluid which regulates body temperature. The pupils of the eyes dilate to receive more light and the face takes on expressions of fear. These include a furrowing of the brows, raising the upper lip and stretching out the lips. Body hair straightens (stands on end). When we are confronted with a potentially life-threatening situation, we adopt a protective posture by cowering to cover the head and upper regions of the body with our hands and arms.

Why do we blush when nervous or embarrassed?

We blush when we are involved in a situation which appears to show that we have transgressed the normal rules of our culture, and created a poor impression of ourselves. For example, in Western culture, it is thought impolite to fart or burp in public. The flushed appearance is an acknowledgement of our responsibility, and a visual act of appeasement to the group. When attention is brought to our discomfort by others, the embarrassment and blushing becomes worse. We also tend to hunch our shoulders to make us look smaller. This posture is also adopted by other mammals in an appeasement situation. Sometimes, feelings of inadequacy will also produce blushing, such as when we are close to another person we consider sexually attractive or intellectually superior. The embarrassment of making a speech to a large audience can also turn our cheeks very red.

The skin is supplied with blood and nerves through a very complex network of vessels. Blushing is caused by the dilation of blood vessels. The face and neck have more vessels per unit volume than other parts of the skin, which is why it is more noticeable in these areas.[24]

What is stress?

The anthropologist Robert Sapolsky says that zebras handle stress better than human beings.[25] This is because their stress is short-lived, based on an immediate emergency situation. They either escape from predators or get eaten. Either way, their stress does not last long. Humans, on the other hand, become stressed when they lose control of an *ongoing* situation. It can be caused by a variety of reasons such as pressure to perform well at work or in sports. Worrying about the lack of money is another common cause of stress, as is moving home, divorce, bereavement or unemployment.

The symptoms of stress also differ widely. They include being depressed, over-tired, irritable, losing sex drive or increasing usual levels of drinking alcohol or smoking. There can also be a loss of appetite or the opposite effect – comfort eating. Stress results from an interaction between the immune system which controls our ability to fight infection, the nervous system which controls our physiological responses, and the endocrine system which controls hormonal activity. Figure 6.2 illustrates how these interactions affect each system.

Figure 6.2 Stress-related interactions between Nervous, Immune and Endocrine Systems

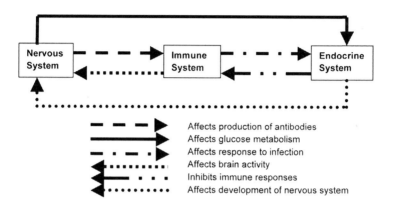

The stress response is controlled by the hypothalamic-pituitary-adrenal axis (HPA axis) and the sympathetic nervous system. Under medium- or long-term stressful conditions, the hypothalamus in the brain produces corticotrophic-releasing factor (CRF). This binds to receptors in the pituitary gland to produce the adrenocortico-tropic hormone (ACTH). This is transported to the adrenal glands, which is stimulated to produce the stress hormone cortisol. Cortisol accumulates in the body, and produces physical effects such as heart palpitations, excessive tiredness or skin problems.

The adrenal glands secrete the hormone epinephrine (also known as adrenaline). The neurotransmitter norepinephrin (also known as noradrenaline) is secreted by the hypothalamus and other areas in the frontal cortex. Their combined action causes increases in heart rate triggering the release of glucose, and the muscles are supplied with an increased blood supply for energy. This makes them ready for exercise in emergency situations. However, when the stress is ongoing, and does not need the fight-or-flight type of response, the muscles remain tense and cause aches to the neck and back and tension headaches.

It has been suggested that most of the harmful effects of stress result from long-lasting corticosteroid secretions. These lower the activity of natural killer cells in the immune system. This means we are more likely, more often, to get colds and other infections.[26]

Stress has been linked to the risk of cardiovascular disease, caused by high blood pressure and high levels of blood cholesterol. One study suggested that people could be grouped into two kinds of different personality.[27] Type A people are impatient, aggressive and react quickly in both movement and speech. They have low self esteem and cannot achieve their own personal goals. They are likely to smoke and drink and have a higher risk of heart disease. Type B people are generally more tolerant and good-natured, and more able to cope under stressful conditions. Most of us have a mix of both types of personality, with one being more dominant.

Why does stress affect digestion?

When we are stressed, blood vessels to the digestive system and other areas constrict, and as mentioned before, the blood is redirected towards the muscles in preparation for handling the crisis. There is a reduced level of saliva in the mouth which is why our throats feel dry when we are frightened. Bowel and bladder are encouraged to remove waste and stop further activity in these last parts of the digestive system. Abnormal hormone secretions also mean the system is not functioning at optimal performance levels. The large amounts of the hormone CRF produced by the hypothalamus reduces appetite, but the steroids triggered by CRF can also produce feelings of hunger. This is why some people lose their appetite when stressed while others comfort-eat large quantities of sugary or carbohydrate foods. The result may include stomach aches, nausea or diarrhoea. Long-term stress can also result in chronic conditions such as Irritable Bowel Syndrome (IBS), indigestion or ulcers. This may be due to reduced immune reactions that can enable infections to flourish at various places along the digestive tract.

How does Post-Traumatic Stress Disorder affect the body?

Following a traumatic event, some people experience symptoms of stress. These include depression, reliving the event, nightmares, loss of appetite, sleeplessness, irritability, and forgetfulness. Nervous reactions, such as the startle response at a loud noise can also be extra sensitive. It seems that high levels of stress hormones such as serotonin continue to invade the body and over-activate areas of the brain involved with long-term memory. A cycle of remembering the event is set up, which reinforces the physiological effects of the stress.

REFERENCES

[1] James, W., (1884), What is an emotion? *Mind*, 9, pp. 188-205 and (1890) *Principles of Psychology*, New York, Holt. Lange, C.G. (1887), Über Gemüthsbewegungen, Leipzig, East Germany: T.Thomas, 1887.

[2] Cannon, W.B., (1929), Bodily changes in pain, hunger, fear and rage. New York: Appleton

[3] Cannon, W.B., (1927), The James-Lange Theory of emotions: a critical examination and an alternative theory. *American Journal of Psychology*, 39, pp. 106-124

[4] Schachter, S., and Singer, J.E., (1962), Cognitive, social and physiological determinants of emotional state. *Psychological Review*, 69, pp. 379-399

[5] Panksepp, J., (1995), Affective neuroscience: a paradigm to study the animate circuits for human emotions, in *Emotion: Interdisciplinary Perspectives* (eds. R.D. Kavanaugh, B. Zimmerberg and S. Fein), Lawrence Erlbaum, Mahwah, pp. 29-60

[6] Mackintosh, B., (2004), Emotions and Mind, Bk 6 Biological Psychology: Exploring the Brain, Open University, Milton Keynes, p.73

[7] Plutchik, R., (1994), *The Psychology and Biology of Emotion*, New York, HarperCollins

[8] Darwin, Charles, (1872/1965) *The Expression of the Emotions in Man and Animals*, Chicago, Chicago University Press

[9] Field, T., Woodson, R.., Greenberg, R., and Cohen, D., (1982), Discrimination and imitation of facial expressions in neonates. *Science*, **218**, pp. 179-181

[10] Erikson, E., (1950), *Childhood and Society*, New York, Norton

[11] Lewis, M., (2000), The emergence of human emotions. In M. Lewis and J. M. Haviland-Jones (Eds), *Handbook of Emotions* (2nd edn.) New York, Guilford, pp. 265-280

[12] Papez, J.W., (1937), A proposed mechanism of emotion. *Archives of Neurology and Psychiatry*, **38**, pp. 725-745

[13] MacLean, P.D., (1949), Psychosomatic disease and the "visceral brain": Recent developments bearing on the Papez theory of emotion. *Psychosomatic Medicine*, **11**, pp. 338-353

[14] Aggleton, J.P., and Mishkin, M., (1986). The amygdala: sensory gateway to the emotions. In *Emotion – Theory, Research and Experience*. Vol.3. *Biological Foundations of Emotion* (eds. R. Plutchik and H. Kellerman), Orlando, Academic Press, pp. 281-299

[15] Macmillan, M.B., (1986, A Wonderful Journey through skull and brains: the travels of Mr. Gage's tamping iron. *Brain and Cognition*, **5**, pp. 67-107

[16] Fridland, A., (1988) What can asymmetry and laterality in EMG tell us about the face and brain? *International Journal of Neuroscience*, **39**, pp. 53-69

[17] Heller, W., (1990), The neuropsychology of emotion: developmental patterns and implications for psychopathology. In *Psychological and Biological Approaches to Emotion* (eds. N. L. Stein, B. Leventhal and T. Trabasso), Hillsdale, Lawrence Erlbaum, pp. 167-211

[18] Dabbs, J.M., Ruback, J.M., Frady, R.L., and Hopper, C.H. (1988) Saliva

testosterone and criminal violence among women. *Personality and Individual Differences*, **9**, pp. 269-275

[19] Mazur, A., and Booth, A., (1998), Testosterone and dominance in men. *Behavioural and Brain Sciences*, **21**, pp. 353-397

[20] Aronson, E., Wilson, T.D., and Akert, A.M., (2005), *Social Psychology* (5th edn.), New Jersey, Prentice Hall

[21] Frey, W. H., (1985), *Crying: The Mystery of Tears*. Minnesota, Winston Press

[22] Derks, P., (1992), "Category and Ratio Scaling of Sexual and Innocent Cartoons", *Humor, International Journal of Humor Research*, **5.4**, pp. 319-330

[23] Berk, I.S., Tan, S.A., Fry, W.F., Napier, B.J., Lee, J.W., Hubbard, R.W., ewis, J.E., and Eby, W.C., (1989), "Neuroendocrine and stress hormone changes during mirthful laughter". *The American Journal of the Medical Sciences*, **298**, pp.390-396

[24] Wilkin, J.K. (1988), Why is flushing limited to a mostly facial cutaneous distribution? *J Am Acad Dermatol* **19**, pp. 309-313,

[25] Sapolsky, R.M., (2004). *Why Zebras Don't Get Ulcers*. (3rd Edn.), Henry Holt & Co.

[26] Selye, H (1976), The Stress of Life, New York, McGraw-Hill

[27] Friedman, M., and Rosenman, R.H.,(1959) Association of specific overt behavior patterns with blood and cardiovascular findings – Blood cholesterol level, blood clotting time, incidence of arcus senilis and clinical coronary artery disease. *Journal of the American Medical Association*, **162, 99**. 1286-1296

CHAPTER 7

HEARING

What is sound?

How do we hear?

How does the ear transmit sound?

Can the unborn baby hear?

Why do we feel giddy and unbalanced when we twirl?

How does the brain convert air-waves into meaningful sounds?

How does the brain interpret music?

Why does too much noise make us feel ill?

What causes hearing loss?

What is ringing in the ears?

Why do some people 'see' sounds as colours?

What is sound?

Sounds are vibrations or air molecules that travel through the air caused by speech or music or any other noise, displacing the medium it is in. This can be any form of matter such as a liquid, gas, or solid. Sound moves along waves of alternating pressure at various frequencies. Humans absorb sound mainly through the ears, although some vibrations can be felt through the sense of touch.

The speed of a sound wave varies depending on whether it is travelling through a liquid, gas or solid. Technically, the speed of sound is proportional to the square root of the ratio of the degree of rigidity, and also to the density of the medium. Sound waves create pressure, which alters the surrounding pressure, and is measured in pascals.

There are two main classes of sound: irregular and periodic. Irregular, random pressure changes are noise-like, for example the sound of a waterfall or the sound of the letter 's' or 'f'. The periodic class has regularly repeated sounds which are higher pitched. They have complex tones such as musical notes. The frequency is the number of pure tone repetition rates.[1]

A sine wave is an unbroken wave with a steady frequency and amplitude. Periodic complex sound is made up of sine waves with different frequencies and at different levels. The frequency response of amplifiers and loudspeakers are based on studies, which measure how different sine waves behave. An amplifier will exactly reproduce a single sine wave or a complex mix of sine waves[2]. High-frequency sounds will be heard as a sharp noise, and low-frequency sounds are much duller and softer.

Sounds can vary in pitch, loudness or timbre. The pitch is the frequency of the vibration, and is measured in Herz (Hz) or cycles per second. Loudness, also known as amplitude is the intensity of changes in the sound waves. More intense sounds have stronger vibrations. Timbre is the type of sound heard. These will be discussed more fully in the next section.

How do we hear?

The human ear is divided into three main sections: the outer ear, the middle ear and the inner ear. Each has a different function in transferring a chain of information on the sound vibrations it receives. These will ultimately be interpreted by the brain as an identifiable noise. Figure 7.1 details the different parts of the ear.

The function of the outer ear is to capture sound. It consists of the external ear or pinna, and is shaped like a convoluted cup. Changes in air pressure are directed through the auditory canal to the ear-drum or tympanic membrane, and cause it to vibrate. The shape of the external ear amplifies sound in speech frequencies ranging from 5 to 20 decibels, and improves the signal-to-noise ratio for speech sounds.[3]

The ear-drum covers a cavity in the middle ear, which contains three tiny bones, known collectively as the auditory ossicles. These are the malleus, the incus and the stapes, also known as the hammer, the anvil and the stirrup, because they roughly resemble these shapes. The main function of the middle ear is to transfer the vibrations from the ear-drum to the inner ear

Figure 7.1 Anatomy of the human ear

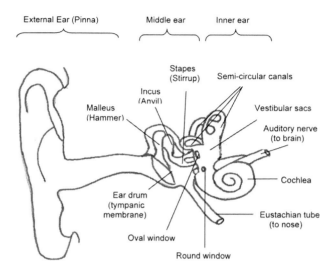

via the three ossicles. The sound waves transfer to the malleus, which in turn moves the incus and pass on the vibrations to the stapes. The stapes footplate presses against another membrane called the oval window. Below this is the round window membrane. It is more flexible, and operates in an opposite direction to the movement of the oval window. When the stapes moves, the round window allows the fluid inside the cochlea of the inner ear to be displaced.

Inside the inner ear is a fantastic structure looking like a snail shell with three loops on top. The coiled arrangement is called the cochlea, and it is filled with fluid. Along its curled length is the basilar membrane, which moves at different locations depending on the frequency of the pressure waves it receives from the oval window.

The organ of Corti consists of the basilar membrane, receptor hair cells and the tectorial membrane. The basilar membrane moves with the tectorial membrane and shifts the hair cells, to release neurotransmitters. These transmit information along the auditory pathway to the brain, by producing electrical activity. This passes along the nerve cells. A neurotransmitter is a chemical released from a nerve fibre.

Figure 7.2 shows the frequency levels measured in Herz (Hz). The cochlea detects high frequency pitch sounds by place coding and low frequencies by rate coding. Low-frequency sounds activate the inner part of the coil, with the lowest at the apex. High-frequency sounds activate the outer part of the coil starting at the base next to the oval window membrane. Sounds with many frequencies cause simultaneous movements along the length of the basilar membrane.

The cochlea is very sensitive to sound. The softest sounds move the tip of the hair cells at between 1 and 100 picometres. A picometre is a million millionths of a metre. The loudest sounds move the tips of the hair cells at 100 nano-metres, that is one thousand millionths of a metre.[4] Messages are sent to the brain about the loudness of a sound by altering the rate of electrical activity in the neurons. Intense vibrations cause the release of more neurotransmitter at the tips of the auditory hair cells thereby increasing the firing rate.

Most sounds we hear have a complex mix of frequencies and timbre. We can distinguish different musical instruments by their timbre. The waveform repeats at a frequency related to the pitch of the tone, as well as overtones, which multiply the frequency at different intensities. Each sound is uniquely coded, for interpretation by the brain via the auditory pathway.

Figure 7.2 Diagram of cochlea showing where frequency sounds are activated.

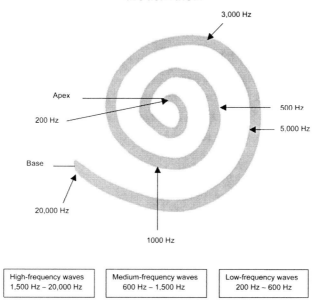

Can the unborn baby hear?

There is evidence that after 24 weeks, the foetus responds to external sounds, such as maternal voice and heart-beat.[5] The mother's voice reaches the unborn baby through her own body as well as externally, and is therefore stronger than other sounds.[6] Research has shown there is a five-second stimulus, which can cause changes in foetal heart rate and movement and last for up to one hour. Another study has discovered that behavioural responses are noted at 16 weeks into pregnancy, whereas the ear is not fully formed until eight weeks later.[7] This indicates that a primitive method of hearing begins with the skin and skeleton until the ear is fully functional. They also found that from 19 weeks, the foetus responds to lower frequencies best, and higher tones as the baby develops.

Why do we feel giddy and unbalanced when we twirl?

At the top end of the cochlea lie two vestibular sacs, which react to gravitational force, and above them the three semi-circular canals. Together, they form the vestibular system. Its main function is connected with balance, and enabling the head to stay in an upright position. Each sac has tissue with receptive hair cells. When the head changes its orientation, these cells are activated[8]. Stimulation of the sacs can cause nausea, as in travel sickness. Fluid inside the canals moves when the head shifts position. This movement causes minute hair cells to release neurotransmitters and promote electrical activity along the nerves, to the cerebellum in the brain (see figure 7.3), which analyses and interprets the information received. Each semi-circular canal lies at a different angle, corresponding to the major planes of the head: sagittal (back to front), transverse (left to right) and horizontal. They respond to the unsteady rotation of the head, and stimulation can cause dizziness.[9] Children like to play the game of twirling around as fast as they can to produce a giddy unbalanced feeling. It also explains the feelings we get from fun-fair carousels.

How does the brain convert air-waves into meaningful sounds?

Figure 7.3 shows the auditory cortex situated in the temporal lobe of the brain. The auditory system can distinguish the direction of a sound. This is because the sound travels to the nearest ear slightly earlier than to the other ear, starting the electrical activity in the neurons sooner. This difference in time enables the brain to interpret where the sound is coming from.

Figure 7.3 Auditory cortex in temporal lobe

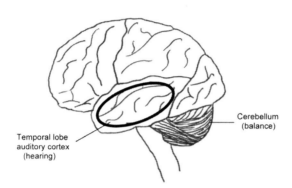

The cochlea nerve transmits information from the organ of Corti. Figure 7.4 shows the main pathways to the brain. The main route is to the thalamus and on to the auditory cortex in the temporal lobe. Another pathway is from the thalamus to the cortex and then the amygdala. A third route is directly from the thalamus to the amygdala. This short-cut is taken at times of danger or threat. The amygdala is an organ in the brain processing emotional information, and enables a person to act quickly in a hazardous situation, and to sense whether we should run fast away from danger, or stay and fight it out if the odds appear to be in our favour.

The auditory cortex also transmits information through descending pathways, which adjusts the sensitivity of sounds and provides a feedback system to the cochlea. It serves to increase the threshold of the signal-to-noise ratio and reduces the risk of damage to the delicate hair cells.[10]

Figure 7.4 Auditory pathway for each ear

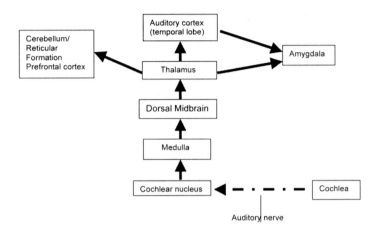

How does the brain interpret music?

Figure 7.5 illustrates that the auditory cortex is divided into primary, secondary and tertiary sections. Cells in the primary auditory cortex are sensitive to specific frequencies. This is called tonotopic organisation, and identifies the pitch and volume of music.[11] It receives its input directly from the thalamus. The secondary auditory cortex seems to process melody and rhythm. Finally the tertiary auditory cortex integrates the information from the other two.

It seems that the right side of the cortex is more responsible for perceiving pitch, melody, harmony, timbre and rhythm. The left side processes changes in frequency and intensity. Both sides of the brain are needed however, to correctly interpret rhythm. The frontal cortex also plays a part in understanding melody and rhythm. It has been noted that the motor cortex is also implicated, for example, foot-tapping or finger-waving in time with the music. Some musicians also make facial expressions when playing music. The areas of the brain linked to emotional responses are also involved, which is why music can make us feel happy or sad.[12] Complicated rhythms utilise additional parts of the cortex and cerebellum. Studies have shown that a fast song can increase the rate of heart beat, and a slow song can slow it down.[13]

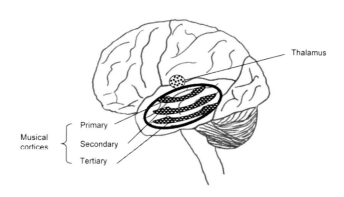

Figure 7.5 Musical cortices area

Why does too much noise make us feel ill?

Too much noise can result in hearing loss, as well as other physical and psychological effects. Environmental noise from road traffic, aeroplanes, industrial activity and loud music all contribute to these ill effects. Also, the external and middle ear greatly amplify sound levels as the sound travels to the cochlea in the inner ear. Exposure to excessive noise levels damages the hair cells irretrievably, creating loss of hearing. This is especially acute in industrialised countries.

High noise levels release extra amounts of the hormone

adrenaline, and this can constrict blood vessels. Prolonged exposure can lead to high blood pressure and ultimately heart disease. Studies have shown that in early pregnancy foetal development can also be adversely affected.[14] Other stress-related effects can result in headaches, tiredness, anxiety and increased aggressive behaviour.

What causes hearing loss?

Hearing loss is defined by the degree of loudness which must be reached before a sound is heard. This is measured in decibels (dB). There are different levels of hearing loss.

Mild hearing loss is in the range of between 25 and 40 dB in adults. People who suffer this degree of loss will find difficulty keeping up with conversations, especially in a noisy environment, such as at parties or restaurants.

Moderate hearing loss is in the range of between 41 and 55 dB. A hearing aid would be an advantage, especially in certain difficult situations.

Severe and profound hearing loss occurs in the ranges of between 71 and 90 dB. People who suffer this degree of loss would need to lip-read and use sign language. Cochlear implants are being developed which are helping people with these levels of impairment.

There are two types of hearing impairment: conductive and sensorineural. Conductive hearing loss occurs when the sound is not conducted through the outer or middle ear. This can happen when there is excessive wax in the ear, or an obstruction enters the external ear canal and blocks direct access to the ear-drum.

Sensorineural hearing loss is caused by damage to the tiny hair cells in the cochlea. The impairment can vary depending on the degree of damage. This can be caused by a noisy environment, or exposure to certain diseases, such as measles, mumps, meningitis or multiple sclerosis. Some drugs such as aspirin or the antibiotic streptomycin can also harm the ear.

There can also be inherited causes of hearing loss. A family with a dominant gene for deafness can pass this on to future generations. Also babies born prematurely may be partially deaf, due to immature development.[15]

What causes ringing in the ears?

Tinnitus is the name given to a disorder whereby a sound is heard in one or both ears, without an external stimulus. It is often described as a continuous ringing or buzzing noise, and is very disconcerting. It is difficult to measure, because it is a subjective noise with no obvious signal to the ear. Some effects can be objectively considered such as an increased blood flow near the ear.

Tinnitus may result from a variety of causes, usually associated with the sensorineural system in the inner ear and brainstem. People who work in very noisy environments are especially at risk. Other causes include Ménière's disease, in which there is increased pressure in the inner ear and otosclerosis, whereby the auditory ossicles become fixed. Ear infections, wax and 'glue ear' in children are also common causes. High doses of aspirin, quinine and some antibiotics can also contribute to this condition. People who are anaemic or who have high blood pressure can become affected. Tinnitus can also result as an after-effect of a head injury following an accident. Sufferers from depression and anxiety psychosis can also complain of symptoms.

In the absence of any obvious cause, it is difficult to explain why people have subjective tinnitus. Recent studies indicate that head and neck nerves, which enter the brain near the auditory cortex, may confuse incoming signals.[16] Treatments vary depending on the overlying causes. Sometimes unexplained tinnitus will disappear without treatment.

Why do some people 'see' sounds as colours?

Synaesthesia is a neurological condition joining two or more sensations that are usually experienced separately. People with this condition may imagine a specific colour when hearing a particular musical note or pitch. The most common form is where colours relate to certain letters, words or numbers, but can also apply to smells, tastes or shapes. Synaesthesia is automatic and involuntary, and occurs in people with normal brain function.

People with music-colour synaesthesia experience colours in response to tones, timbre or key. Individuals are consistent in their choice of colour relative to the musical experience, but there is variation between people with this condition. There is disparity also in the hue of the colour, and the colours can move in and out of field of vision.

Synaesthesia appears to be genetic. It runs in families, and affects slightly more women than men. Research suggests a link along the female line. For example, father-to-daughter, mother-to-son, mother-to-daughter transmissions are common. Members of a synaesthetic family can have different types of the condition, which seems to show that developmental factors also play a part.

Different regions of the brain specialise in specific functions, such as vision, hearing and language. One theory to explain synaethesia is that there is a "cross-activation" in these areas in people with the condition. For example, the region involved in processing colours is adjacent to the region connected with understanding sounds. Another theory suggests that the normal levels of neuronal activity become unbalanced and heighten activity in these areas. This could explain why psychedelic drug users often report synaesthetic experiences.

It is also thought to be a condition of creative people. Composers who have reported the condition include Duke Ellington who had timbre- colour synaesthesia and Franz Liszt who had music- colour synaesthesia.[17]

REFERENCES

[1] Gregory, R.L. (Ed.), and Zangwill, O.L. (1987) *The Oxford Companion to The Mind*, Hearing, Oxford University Press, pp. 303-308

[2] Békésy, G. von (1960) *Experiments in Hearing* (trans. and ed. By E.G. Wever). New York and Moore, B.C.J. (1982) *Introduction to the Psychology of Hearing*, 2nd edn. London in Gregory, R.L. (1987) *The Oxford Companion to The Mind*, op. cit.

[3] Brugge, J. F., and Howard, M.A., (2002) Hearing, chapter in *Encyclopedia of the Human Brain*, ISBN 0122272102, Elsevier, pp 429-448

[4] Carlson, N. R., (1994) *Physiology of Behaviour*, Fifth Edn., Allyn and Bacon (Paramount Publishing), p.193

[5] Fifer, W.P. and Moon, C. (1988) Auditory experience in the fetus in *Behaviour of the Fetus* (eds. W.P.Smotherman and S.R. Robinson) The Telford Press, Caldwell, pp.175-188 in Toates, F., (2001), *Biological Psychology*, Pearson Education Limited, p.223

[6] Busnel, Granier-Deferre, and Lecanuet (1992) in Chamberlain, D.B., Life Before Birth, The Fetal Senses,
http://www.birthpsychology.com.lifebefore/fetalsense.html.

[7] Shahidulla, S. and Hepper, P. G. (1992) Hearing in the Fetus: Prenatal Detection of Deafness. *International Journal. of Prenatal and Perinatal Studies* **4** (3/4), pp 235-240 in Chamberlain, D.B., op. cit

[8] Toates, F., (2001), *Biological Psychology*, op. cit. pp.224-5

[9] Dickman, D., Vestibular System Primer,
http://vestibular, wustl.edu/vestibular.html

[10] Toates, F., (2001) *Biological Psychology*, op. cit. pp 222-223

[11] Arlinger, S., Elberling, c., Bak, C., Kofoed, B., Lebech, J., Saemark, K., (1982), Cortical magnetic fields evoked by frequency glides of a continuous tone. *EEG & Clinical Neurophysiology,* **54,** pp. 642-653

[12] Cromie, W.J., Music on the brain: Researchers explore the biology of music in Harvard University Gazette
http://www.hno.harvard.edu/gazette/2001/03.22/04-music.html

[13] Bernardi, L., Porta, C., Sleight, P., (2006) Cardiovascular, cerebrovascular, and respiratory changes induced by different types of music in musicians and non-musicians: the importance of silence, *Heart,* **92,** pp. 445-452

[13] *Noise: A Health Problem*, United States Environmental Agency, Office of Noise Abatement and Control, Washington, DC 20460, August 1978

[14] Gregory, R.L.(Ed) *The Oxford Companion to The Mind*, op. cit. pp.307-8

[15] http://ohsu.edu/ohrc/tinnitusclinic/compTreatments.html

[16] Greenfield, S., (2000), *The Human Brain, A Guided Tour*, The Guernsey Press Co., Ltd., Guernsey, C.I., pp 67-8 and BUPA's Health Information Team, (Oct. 2003) Tinnitus.
http://hcd2.bupa.co.uk/fact_sheets?Mosby_factsheets/Tinnitus.html

CHAPTER 8

HORMONES (ENDOCRINE SYSTEM)

What are hormones?
What does the endocrine system consist of?
Which parts of the brain are involved in the endocrine system?
Disorders affecting the endocrine system

What are hormones?

The endocrine system produces internal secretions of hormones. Exocrine glands secrete fluids through ducts to the outside of the body, such as tear glands, sweat glands or salivary glands. The endocrine system is ductless, and the hormones produced by the glands act as chemical messengers in the bloodstream to specific target cells throughout the body. When they arrive at their destination, they bind to receptor cells and change their function to meet environmental demands and restore homeostatic balance. For example, during exercise the body needs additional oxygen and nutrients to be sent to the muscles and heart, and hormones support other structures to produce more. When the body rests, hormones inhibit the production of excess oxygen and nutrients to meet the reduced demand. Hormones regulate and affect many different aspects of the functions of the body including growth, metabolism, behaviour, the immune system and reproduction. Metabolism is the complete set of chemical reactions in living cells that are essential for growth and reproduction, maintenance of body systems and responses to the outside world.

There are three main types of glandular hormone: protein, amine and steroid. Protein hormones are long strings of amino acids. Short strings are called peptide hormones. Examples are adrenaline and insulin. Their receptors are situated on the walls of cells. These in turn produce secondary messengers which actively alter the function of the cell. Amine hormones are derived from a single amino acid. An example is thyroxin. Steroid hormones are made from rings of carbon atoms. They can become lipids, which are fatty acid molecules. Their receptors are inside the cell and bind to DNA, which affects genes. Example of sex steroids are testosterone and progesterone.

Another type of hormone is produced not by glands but by specific body tissues. These local hormones act only in the immediate vicinity. They include histamine which is released when an area of the body is damaged. Histamine produces the inflammation which surrounds a wound, by causing the blood vessels to dilate to increase the blood supply to the area. Serotonin is another hormone found in the platelets of the blood. It is also produced in the brain, where it manipulates behaviour. However, most serotonin is synthesised in the intestinal tract, where it influences gastric secretions. Other hormones in the intestines, such as gastrin and cholecystokinin (CCK) also affect secretions of digestive juices.

What does the endocrine system consist of?

Figure 8.1 shows the main organs that produce hormones.

The pineal gland is also known as the pineal body or the epiphysis. It is pea-sized and shaped like a pine cone, located just above the brainstem. It releases the hormone melatonin. Secretions are inhibited by daylight, and stimulated at night. Melatonin appears to regulate our biological clock cycle, and can affect sleep-wake patterns. Melatonin is produced at much higher levels before puberty, when it prevents the growth and development of reproductive organs in children. After puberty the gland tends to shrivel and harden.

Figure 8.1 Main organs of the Endocrine System

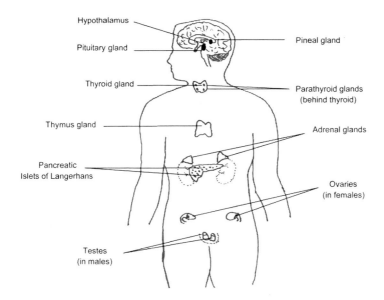

The pituitary gland is small and oval shaped. It consists of two parts that have very different functions, the anterior and posterior pituitary. The anterior (frontal) pituitary produces several hormones that influence other glands: Adrenocorticotrophic (ACTH) hormone targets the outer cortex of the adrenal glands, to influence secretion of cholesterol and steroids. The output of ACTH also has a circadian cycle, its maximum production being early morning, and lowest at the end of the day.

Thyrotrophin (TSH) targets the thyroid gland, and affects its secretions.

FSH is a gonadotrophin (sex cell stimulating) hormone that activates the growth of follicles in the female ovaries to produce oestrogen and also excites male testes to produce sperm.

LH is a gonadotrophin luteinizing hormone which stimulates the ovaries to produce progesterone. It also promotes the development of interstitial cell stimulating hormone (ICSH) to support the production of testosterone from the testes.

Growth hormone (GH) targets the liver, intestines and pancreas, where it regulates metabolism by synthesising proteins, and increasing blood glucose levels. It also breaks down adipose (fatty) tissue. GH is mainly produced at night during sleep. It is most active during childhood and stimulates the growth of bones and skeletal muscles.

As its name suggests, prolactin targets the mammary glands of the female breasts and stimulates milk production after childbirth.

The posterior pituitary produces two hormones that have very different effects on the body: antidiuretic hormone (ADH), often called vasopressin (AVP), and oxytocin hormone.

ADH conserves water by slowing down the production of urine as required. For example, if the body is dehydrated ADH allows more water to be reabsorbed. It binds to receptor cells on the kidneys, which are activated to allow the transport of water through kidney tubules and back into the blood stream. When the body has received a large intake of fluid, ADH production is inhibited and water cannot permeate through the kidney membranes. Consequently, urine is formed and passes out of the system. When blood pressure is low, for example following a haemorrhage, ADH production increases and causes tiny blood vessels to contract, thereby raising blood pressure.

Oxytocin stimulates the muscles of the walls of the uterus to contract, in response to the stretching of the cervix during childbirth. The baby's head is forced downwards, triggering the release of greater levels of oxytocin and more frequent muscular contractions. After birth, the hormone triggers contraction of cells in the mammary glands which store milk. A conditioned response to the baby crying can stimulate oxytocin secretions in the mother.

The thyroid gland is shaped like a butterfly in the neck. It produces three hormones: thyroxine (T_4), triodothyronine (T_3) and calcitonin.

T_3 and T_4 need iodine which is obtained from food. Diets deficient in iodine create an enlarged thyroid. The iodine combines with tyrosine to form T_3 and T_4. Secretion of these hormones is regulated by TSH hormone from the anterior pituitary, which itself is stimulated by thyroid releasing hormone (TRH) from the hypothalamus in the brain. The thyroid hormones are involved in almost every aspect of our body functions. They affect metabolism by regulating the synthesis of proteins, fats and carbohydrates, and are essential for normal development.

Calcitonin regulates levels of calcium and phosphorus in the body. It prevents calcium being reabsorbed from the bones and into the blood. This is done by inhibiting the activity of cells called osteoclasts which digest bone matrix, and release calcium and phosphorus into the blood stream. In the kidneys, calcitonin slows down their re-absorption and is ultimately channelled out of the body in urine.

There are four parathyroid glands situated at the back of the thyroid gland. They release parathyroid hormone (PTH), which, unlike calcitonin, is released when levels of calcium and phosphorus in the blood are low, thereby increasing levels. It has therefore opposite effects on the bones and kidneys to the thyroid hormone.

The thymus gland has two lobes which contain lymphoid cells called thymocytes. It secretes a group of hormones called thymosins, which stimulate the production of the T-lymphocyte white blood cells (the 'T' stands for thymus). The gland forms part of the immune system and protects the body against unwanted substances (see chapter on Blood Circulation). The thymus gland is active from birth until puberty and thereafter starts to wither due to the increased circulation of sex hormones.

The adrenal glands rest on the kidneys. Each one consists of two parts: the adrenal cortex and the adrenal medulla. The adrenal cortex is the outer and largest part (about 80%) surrounding the inner medulla. It produces three types of steroid hormones in response to stimulation from ACTH from the pituitary gland: mineralocorticoids, glucocorticoids and androgens (sex hormones).

Mineralocorticoids maintain a balance between fluids and different minerals, especially potassium and sodium in the body. The main hormone is aldosterone, which is released when potassium levels in the blood are high. Aldosterone acts on the kidneys to retain sodium and conserve water by reducing urination. It also allows the release of excess potassium into urine to be passed out of the body.

Glucocorticoids regulate the metabolism of glucose, fats and proteins. The main hormone is cortisol (see chapter on Emotions and Stress). It is vital for the maintenance of essential body functions. Cortisol responds to stress by increasing blood pressure and blood sugar levels. A synthetic form called hydrocortisone treats allergies and inflammation. The breakdown of fats and proteins create energy. Cortisol also suppresses immune reactions, which can be useful in treating autoimmune diseases.

There are tiny amounts of sex hormones in the adrenal cortex, which influence adult patterns of body hair in males and females.

The inner core of the gland, the adrenal medulla produces the hormones adrenaline and noradrenaline, which respond to signals from the nervous system to deal with stressful situations. Adrenaline increases heart rate and blood pressure and noradrenaline dilates blood vessels to allow more blood to reach the heart, brain and muscles. Smaller blood vessels are constricted, which increases the metabolic rate and prevents the blood being diverted to less important areas of the body such as the skin. This contributes to us having a paler appearance when we are stressed.

The adrenal medulla, together with the brain enables us to make prompt decisions in times of danger, such as when to run away from a dangerous situation.

The pancreas contains groups of cells called Islets of Langerhans, named after Paul Langerhans who discovered them in 1869. These secrete mainly three types of cell: α (alpha), ß (beta) and δ (delta), the majority of which are beta cells. There are also very small quantities of PP cells which contain polypeptides and ε (epsilon) cells which release the hormone ghrelin. The hormones are secreted directly into the blood stream.

Alpha cells secrete the hormone glucagon which enables the conversion of the animal starch glycogen to glucose in the liver and muscles, when levels are low, for example after exercising. The extra glucose provides energy.

Beta cells secrete the hormone insulin, which lowers increased levels of glucose and other nutrients as well as amino acids and fatty acids. It has the opposite effect to glucagon, by converting glucose to glycogen in the liver and muscles. It stops the breakdown of fats by storing them in adipose tissue.

Delta cells secrete the hormone somatostatin which prevents the secretion of insulin and glucagon.

In summary, insulin activates beta cells and inhibits alpha cells. Glucagon activates alpha cells which in turn activate beta and delta cells. Somatostatin inhibits both alpha and beta cells. The gonads secrete hormones connected with the reproductive system. After puberty, ovaries in females produce the hormones oestrogen and progesterone, and the testes in males produce testosterone. These hormones develop male and female secondary sexual characteristics. In women there is an enlargement of breasts and the start of the menstrual cycle, during which eggs mature ready for fertilisation or are eliminated from the body. There is also an accumulation of fat on the hips and thighs. In men, the voice changes into lower tones, muscles and bones increase size, and sperm and semen is produced. Chest and facial hair also start to grow. In both sexes there is a growth of pubic and underarm hair.

Which parts of the brain are involved in the endocrine system?

The hypothalamus generates releasing and inhibiting hormones that have a direct influence on the secretion of the hormones produced by the pituitary gland. It is connected to the anterior pituitary by the pituitary stalk. The posterior pituitary is really an extended part of the hypothalamus, made up of the axons of neuron cells. Together they regulate many of the hormones in the endocrine system. Figure 8.2 shows the location in the brain and its proximity to the pituitary gland.

Blood passes between the hypothalamus and the anterior pituitary through arteries connected by tiny capillary blood vessels. The hormones are transported through this blood supply, which drains into portal veins between the capillary beds of the hypothalamus and the anterior pituitary, where they meet the cells secreting pituitary hormones.

Figure 8.2 Hypothalamus and pituitary gland in the brain

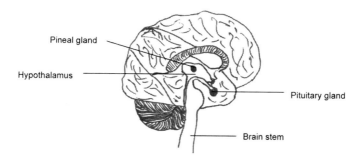

Figure 8.3 Pituitary gland hormones

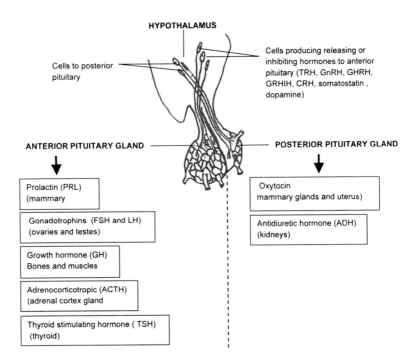

The hormones released by the hypothalamus, and acting on the anterior pituitary gland are as follows:

- Thyrotropin-releasing hormone (TRH), which stimulates the release of TSH and PRL
- Gonadotrophin-releasing hormone (GnRH), which stimulates the release of LH and FSH
- Growth hormone-releasing hormone (GHRH), which stimulates the release of GH, and also GHRIH which inhibits its release
- Corticotropin-releasing hormone (CRH), which stimulates the release of ACTH,

- Somatostatin, which inhibits the release of GH and TSH,
- Dopamine, which inhibits the release of PRL

The posterior pituitary stores hormones which are synthesised in the nerve cells. The cell body contains the nucleus in the hypothalamus, and the cell axons in the pituitary. These are released in short spurts as directed by the hypothalamus. As mentioned previously, the two hormones released by the posterior pituitary gland are ADH (vasopressin) and oxytocin.

Figure 8.3 summarises the effects on the body of hormones controlled by the hypothalamus and secreted by the pituitary gland.

Disorders affecting the endocrine system

Acromegaly – This is caused by too much growth hormone (GH) being released by the anterior pituitary gland in adults. The bones of the hands, feet and lower jaw widen and the brow protrudes, skin thickens and becomes oilier, and the heart can become enlarged. The extra growth can cause an enlargement of the heart, arthritis and numbness of the hands.

Addison's disease – is caused by an underproduction of glucocorticoid and mineralocorticoid hormones secreted by the adrenal glands. The effects are weakened muscles and problems with the gastrointestinal tract, and may be a cause of anorexia. The hormones also affect the menstrual cycle. The disease produces excessive tiredness and chronic dehydration.

Cushing's syndrome – is caused by an overproduction of cortisol secreted by the adrenal glands. The effects are rapid weight gain, especially of the upper body. The skin becomes thin and bruises easily. There can be excessive growth of facial and body hair in women and decreased fertility and libido in men. Blood sugar and pressure levels are high and there is muscular weakness.

Diabetes mellitus – There are two types. Type 1 is caused by the destruction of beta cells in the pancreatic Islets of Langerhans, rendering the remainder unable to produce sufficient quantities of insulin to control the levels of glucose in the blood. The symptoms are excessive thirst and urination. It usually affects children and adolescents, and there are genetic and environmental factors which influence the onset of the disease. Type 2 affects adults and is more common. It is caused by a defective response by the insulin receptor cells in tissues, as well as lifestyle factors such as obesity and lack of regular exercise and also genetic predisposition. Pregnant women also sometimes develop this type of diabetes.

Diabetes insipidus – occurs when the posterior pituitary gland makes insufficient quantities of ADH which leads to excessive production of dilute urine, as water cannot be reabsorbed in the kidneys. The effects are dehydration and high thirst and frequent urination. It is much less common than Diabetes mellitus.

Dwarfism or Lorain Lévi syndrome – The anterior pituitary gland does not produce enough growth hormone GH during childhood resulting in stunted growth.

Gigantism – is caused by excessive secretion of the GH from the anterior pituitary gland during childhood, and results in the abnormal growth of the long bones of the body.

Hyperthyroidism, also known as **thyrotoxicosis** is an overproduction of hormones T_3 and T_4 in the thyroid gland and increases metabolic and heart rate. There is also weight loss with a good appetite, diarrhoea and protruding eyes from excess fatty tissue.

Hypercalcaemia – is an excess of calcium in the blood caused by the overproduction of PTH in the parathyroid glands. It produces abdominal pain, excessive thirst and urination, nausea and vomiting. The condition can also lead to hyperparathyroidism, causing oesteoporosis or brittle bones.

Hypocalcaemia – is a lack of calcium in the blood caused by the under-production of PTH in the parathyroid glands. The effects are numbness and tingling and muscular spasm of the hands and feet.

Hypothyroidism, previously called cretinism (in children) – is caused by an under-active production of the hormones T_3 and T_4 in the thyroid gland. In elderly adults it is also called myxoedema. The effects are abnormally low metabolic rate, lethargy, weight gain or anorexia and constipation. It can be caused by iodine deficiency in the diet. The low levels of T_3 and T_4 stimulates the pituitary gland to release TSH hormone and produce more thyroid tissue, which swells the thyroid and produces a goitre.

Polycystic ovarian syndrome – occurs when there is an abnormally high ratio of the sex hormone LH to FSH, resulting in an overproduction of testosterone in the ovaries. It results in infrequent menstruation, and can cause infertility, obesity and acne.

Seasonal Affective disorder (SAD) – is caused by an overproduction of the hormone melatonin released by the pineal gland. It triggers depression and lethargy during winter, when daylight hours are shorter. There may be a connection to the hibernation instinct in some mammals.

CHAPTER 9

LANGUAGE

How do we talk?
Which parts of the brain are involved in language?
Why can't other primates speak like humans?
How do we recognise the meanings of words?
Why do some people have difficulty
 learning to read and write?
Is there a language gene?
Is language instinctive?

How do we talk?

Figure 9.1 – Vocal apparatus in humans

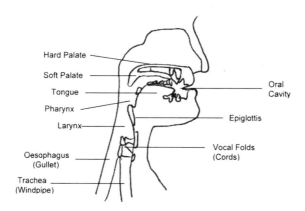

Figure 9.1 shows the main organs that enable humans to speak. Air travels up from the lungs via a tube called the trachea, also known as the windpipe, for obvious reasons, and on to the vocal tract. The larynx is a gateway between the lungs and the mouth, and contains the epiglottis, the vocal cords and the false vocal cords. In males the larynx is seen as the bump in front of the neck called the 'Adam's apple'. Figure 9.2 shows the structure of the larynx. The epiglottis is a flap of cartilage lying behind the tongue and in front of the larynx. During swallowing, this 'gate' is closed, and prevents food and liquids from entering the trachea. Occasionally, minute pieces of food or drink will be misrouted and cause coughing or choking.

Figure 9.2 Structure of larynx

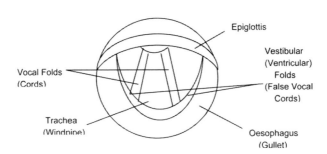

Voice sounds are made by forcing a stream of air from the lungs through the vocal folds, also known as vocal cords. They comprise twin infolds of mucous membrane which stretches across the larynx. As air passes through, the cords vibrate and modulate the flow according to the frequency. Adult men have an average frequency of about 120 Hz, adult females about 220 Hz and in children the frequency is more than 300 Hz.

Men have a larger larynx and larger vocal cords, which measure between 17mm and 25mm in length. Female vocal cords measure between 12.5mm and 17mm in length[1].

The cords are ligaments that attach to the larynx muscles on their outer edges. On either side of the vocal cords is a pair of folds of mucus membrane that stretches across the cavity. These are known as false vocal cords. The scientific name for each is the vestibular fold – sometimes also called the ventricular fold. They protect the delicate true vocal cords, and also help to prevent food from entering the windpipe. These cords are not normally involved in speech production.

The sound made by speech is a pattern of minute vibrations of the air, which vary in frequency distribution. The fundamental frequency band of sound is where the energy is most concentrated and determines pitch. Higher frequency bands are called formant frequencies and indicate the timbre of the sound.

Vowels produce a harmonic sound by periodic vibrations of the vocal cords, causing the molecules of air to move. The pitch of the sound depends on the frequency of the vibration. Changes in shape of the space in front of the vocal cords result in separate vowel sounds. This happens because each vowel uses a different position of lips and tongue relative to the teeth and palate, and the mouth remains partly open. Try sounding out the vowels *a,e,i, o,u*.

Consonants are generally not harmonic, but are scrapes, clicks and bangs, made by closing part of the throat, tongue or lips.[2] Try sounding out the following: *p, t, k, v*.

Using a language is the ability to match sounds with meanings. We learn from a young age that certain sounds always have the same meanings. For example, the word *cat* always relates to a type of animal. We can then pass on this information quickly and easily. Sometimes there are contradictions in the meanings of words, but because we understand the context, we can make sense of the apparent differences. For example, blueberries are berries that are coloured blue, but strawberries are not berries coloured straw.

Words have classes. A noun refers to an object such as *child, book, pen* or *bed*. A verb relates to an action, such as *come, run, ask* or *carry*. Verbal nouns describe the action, such as *arriving, explosion* or *announcing*. Adjectives describe nouns, for example, the *green* parrot, or the *angry* lady. Adverbs modify verbs, such as she came *slowly* towards us or the child *quickly* ate his food. Prepositions and conjunctions link words in a sentence, for example *to, with, on, for*, or *and*. All these types of words are placed in certain orders within a sentence to convey meaning.

There are roughly 7000 languages currently spoken in the world.[3] Each one has a basic set of rules which can be conveyed by speech or writing to other people who are familiar with the system.

Which parts of the brain are involved in language?

Until recently, most of the information on the physiology of language has been obtained from studies of people who have received brain injuries as a result of accident or disease, resulting in speech disorders. Tests have shown that 95% of people who are right-handed use the left hemisphere to produce speech and language, whereas only 70% of left-handed people use this side predominantly.[4] During these tests the left side of the brain is anaesthetised, which causes problems with speech and comprehension, while there is no such effect when the right side of the brain is anaesthetised.

In 1861 the scientist and surgeon Paul Broca was studying a patient who was able only to repeat the word "Tan". Following the patient's death, Broca carried out an autopsy and discovered acute damage to part of the frontal lobe of the brain (see Figure 9.3). It lies next to the area that controls the muscles connected with speech. Further autopsies confirmed

Figure 9.3 Broca's and Wernicke's areas

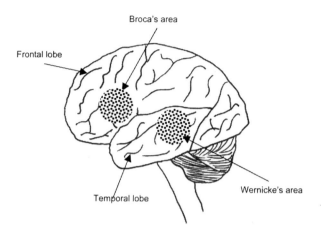

that this area seemed to control the articulation of words. It has since been named Broca's area and people who suffer the disabilities associated with this damage are said to have Broca's aphasia. Their speech is slurred and stilted and they use short sentences. Generally their understanding of what is said to them is good, except where the sentences are complex.

The neurologist and psychiatrist Karl Wernicke subsequently discovered in 1874 another region in the brain that affected a different part of language production. As shown in Figure 9.3, it lies in the temporal lobe and is now known as Wernicke's area. People who have Wernicke's aphasia have fluent speech and grammar, but use inappropriate nouns and main verbs. They have difficulty understanding what is said to them. It seems that this part of the brain provides information about sounds and meanings of words, via auditory codes. Wernicke suggested this is where the memory recalls how sounds are made up into words. More recent evidence from scanning techniques confirms that listening to words increases metabolic activity in this part of the brain.[5]

In 1972, the neurophysiologist Norman Geschwind developed a theory called the Wernicke-Geschwind Model of Aphasia. This shows that language areas are interconnected by a network of neurons. Breaking these connections causes different types of aphasia, depending on where they occur. Figure 9.4 shows the routes taken for hearing and seeing words.

Where the damage lies between the frontal and temporal lobes, patients are unable to repeat words said to them, whilst retaining a good understanding of what is said, and also being able to speak fluently. This condition is called conduction aphasia.

There is evidence that people who are deaf communicate by signing, use the same areas of the brain usually associated with hearing spoken language. Electrical activity in regions of the brain related to receiving visual stimuli is greater in deaf people than those with normal hearing. Studies have shown that the left hemisphere is specialised for receiving symbolic signals, whether they are heard or seen.[6] It would appear therefore, that the left hemisphere controls all linguistic tasks, whether they originate from the ear, the eye, the mouth or the hand.

As mentioned previously, left-handed people do not have the same relationship to the left hemisphere for language, as those who predominantly use the right hand. Whilst the majority use the left hemisphere, in about 19%, the right hemisphere controls language. Management of language appears to be more evenly distributed between both halves of the brain in left-handed people than right-handers. Left-handedness seems to favour mathematical and spatial abilities and also artistic activities, all of which are controlled by the right hemisphere. Interestingly, right-handed people with left-handed relatives appear to differ in the way they analyse sentences.[7]

Figure 9.4– Wernicke-Geschwind connectionist model

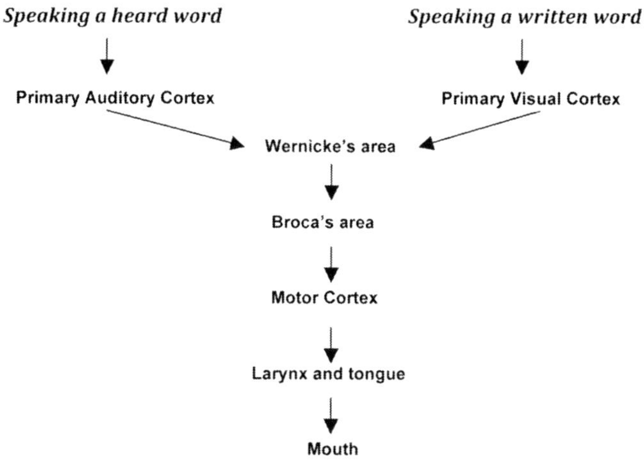

Speaking a heard word → Primary Auditory Cortex → Wernicke's area
Speaking a written word → Primary Visual Cortex → Wernicke's area
→ Broca's area → Motor Cortex → Larynx and tongue → Mouth

Why can't other primates speak like humans?

There have been serious attempts by scientists, carried out over many years to try to make primates talk. Some chimpanzees reared from birth by humans have learned to communicate in fairly complex ways by signing[8]. However, there is an ongoing debate about the true level of linguistic ability in other primates.

Studies carried out watching vervet monkeys in Amboseli National Park in Kenya have shown that they communicate danger to others by a variety of different alarm calls.[9] One sound representing 'watch out there's a snake nearby', causes the group to look down before scampering away. Another sound warning of the approach of an eagle causes the group to look up before scattering. They have one other danger signal signifying a nearby leopard. But that seems to be the sum total of their language skills.

One of the main drawbacks to non-human speech is that the structure of the vocal tract in humans is unique. We are the only species to have the right physiological apparatus to produce complex speech patterns. As can be seen from Figure 9.1, the main vocal structures include the larynx, which houses the vocal cords and the pharynx that opens into the oral and nasal cavities. Most mammals, including apes and newborn humans have the larynx situated high in the neck, thus shortening the pharynx and making it difficult to vary vocal sounds.

The larynx remains in the high position in newborn babies enabling them to suck without choking. It begins to descend from about three months of age, when the baby begins to make language sounds and eat some solid foods, until aged about four years, when the child is able to talk in a fairly intelligible way. The downside for humans is that they are unable to breathe and swallow simultaneously without choking.[10] However, the advantages of complex language skills for communication far outweigh this occasional problem.

How do we recognise the meanings of words and sentences?

The meaning of words is linked in our memory and experience. For example a book has a cover, pages, words and sometimes drawings. A flower is brightly coloured, has buds, petals, leaves, stem, seeds, and often has a perfume. Memories are stored in the association cortex, which connects different regions of the brain.

Sometimes the meaning of a sentence is not clear until it is completed. For example the phrase *'I sat beside the bank...'* could relate to the river bank or the place where money transactions occur. When the sentence is completed by *'... and watched the sun glistening on the water,'* or *'...and waited until it was open for business'*, we understand the correct context.

Damage to the sensory association cortex can affect our ability to identify spatial words such as *up* or *down*, or point to parts of the body. More severe damage to the temporal and parietal lobes can mean people are unable to understand what is being said to them.

Connecting Wernicke's and Broca's areas is a bundle of nerve fibres called the arcuate fasciculus. This conveys information about the sound of words, but not the meaning. Damage to this area causes conduction aphasia, discussed earlier. Patients can repeat sounds they hear, but only if the words have coherent meanings to them. They are unable to repeat sounds that are unrecognisable non-words, such as' *tek* 'or *'phul'*. There is a direct pathway through the arcuate fasciculus. It conveys speech sounds to the frontal lobes that enable us to repeat unfamiliar sounds such as words in a foreign language. Another pathway is indirect and transmits the meanings of words.

Each language has its own set of grammatical rules and word order, which enables us to recognise the meaning. For example *'cat eats bird* 'makes more sense than *'bird eats cat,'* even though the same words are used in the sentence. But a sentence can be grammatically correct and still convey the wrong message. For example, newspaper headlines such as 'Killer sentenced to die for second time in ten years'[11] does not mean he or she was exhumed and killed again. Another headline 'Patient at death's door – Doctors pull him through' obviously doesn't mean medical staff physically manhandled him through the metaphorical door. Our experience tells us which meaning is correct. Some words have more than one meaning, and we can tell from the context of the sentence which one should apply. For example, *branch* can mean a part of a tree or a subsidiary business location; a *ring* may be placed on a finger or a noise made by a telephone or bell; *spring* may be a season of the year or a coiled wire, whereas *sprung* may relate to a mattress or an escape from prison.

Damage to some parts of the brain can lead to the condition called anomic aphasia. People with this condition can comprehend well and speak easily, but they cannot find the right word, and may take a circuitous route to overcome this difficulty. They may not be able to name specific nouns such as people or places. People with damage to the cortex of the temporal lobe cannot recognise other people. In one study, a patient was unable to explain the meaning of words related to living things. He could not name a particular animal, but could describe it accurately.[12] The frontal lobes of the brain are responsible for planning, organising and performing actions.

Damage in this region produces poor verb finding, as is found in Broca's aphasia.

Young children learn words or phrases by connecting them with the context of the situation. For example a child learns to connect seeing a bowl of cereal with the word 'breakfast'. The ability to recognise the meanings of words and learn languages comes easily to children without formal instruction, but gets more difficult as we get older. This can be seen when children easily become bi-lingual, if they are exposed to an environment that regularly uses a second language.

However, most adults find learning a foreign language very difficult. Imaging techniques have shown that language learning in children, including a second language, activates the same brain area. However, adults attempting to learn a second language use different regions of the brain.[13] Adults lose the ability to speak well if they did not learn as children. For example, deaf adults who have regained their hearing are unlikely to learn spoken language.

People use face and hand gestures to reinforce the meaning of their words. It is interesting that people blind from birth use the same hand gestures as sighted people.[14] Also deaf signing utilises the same areas of the brain that hearing people use for speech and listening.

Why do some people have difficulty learning to read and write?

Reading and writing skills are closely allied to listening and talking, and – unlike speech – need to be learned. Much of the information about the way we learn to read and write comes from studying those who have literacy problems. The disorder comes under the generic heading of dyslexia. In the general population, about 5% are dyslexic, with a larger proportion in boys and left-handers. Dyslexia is found in people within all ranges of ability, and is not related to general intelligence.

Post mortem studies on people who were dyslexic but died from unrelated causes, shows unusual groupings of cells on the cortical (outer) surface of the frontal and temporal regions of the brain. These groupings distort the normal layered and columnar arrangements of the cerebral cortex. It is thought that the anomalies may occur during foetal development, when active cells migrate to this region. The results lead to unusual patterns of neuronal connections in the language areas of the temporal cortex.[15]

Functional Magnetic Resonance Imaging (fMRI) scans indicate that dyslexic people have different patterns of brain activity to normal. There is also less activity of the visual cortex and a disrupted spread of activity when responding to written words. Studies have been carried out on a family from Finland with several dyslexic members. Tests showed an abnormality in the gene DYXC1.[16]

Other investigations have shown that there seems to be two different brain systems involved in reading. One focuses on the sounds of letters and the other on the meaning of whole words.[17] Acquired dyslexia (alexia) happens to adults with brain injury or disease, who previously had no problem with reading. Some will read a word incorrectly but with a similar context. For example the word *cow* will be read as *horse* – both are animals seen in fields, but the reader cannot recall the correct name. Some patients cannot recognise similar words that do not sound as they are written, such as cough and plough.

Positron Emission Tomography (PET) scans have indicated that various regions of the brain are activated when attending to words. Passively looking at words on a screen activates the posterior area of the left hemisphere, whereas passively hearing words activates the temporal lobes. Repeating words activates the motor cortex in both hemispheres as well as other areas. Word repetition or reading aloud generates little activity in Broca's area, but when asked to name a verb associated with a particular noun, language related regions in the left hemisphere, including Broca's area were activated.[18]

Is there a language gene?

Genes do not specify how living things behave. Genetic information is stored in the molecule deoxyribonucleic acid (DNA) (see chapter on Atoms, Cells and Genes). It is made up of four different nucleotide bases: adenine (A), cystosine (C), guanine (G) and thymine (T). These bind to the backbone of the molecule. Ribonucleic acid (RNA) is a sugar group ribose in the backbone of the molecule. RNA also contains bases A, C, G and uracil (U). There are long strings of DNA that trigger the copying of proteins and attract neurons into neuronal networks. These are activated during learning, especially in connection with understanding and producing language.

In 1992, a gene called FOXP2 was discovered, which in some forms affects a genetic disorder where the patient has specific grammar defects in speech, and this runs in families[19]. Members of the families without this form of the gene remain unaffected. It has been called the 'language' or 'grammar' gene. Subsequent studies in 2001 showed that the gene has a forkhead binding domain[20]. This codes for a transcription factor, which is a protein that binds to other genes and enables them to make the change from DNA to RNA. The binding should result in the production of an arginine amino acid within the protein. All affected people with the faulty gene produce instead a histidine amino acid in the transcribed protein.

It is unlikely that a single gene can fundamentally be responsible for the range of abilities that are needed to produce language. However, the debate rages on.

Is language instinctive?

There are many arguments amongst linguists and neurophysiologists, and some do believe in the existence of a language instinct.

Babies appear to be born with an innate ability to interpret different sounds. Children seem to develop grammar instinctively.[21] For example, they use the plural 's' in the right places, but which are not in use in our language. For example a young child will pluralise the word *'sheep'* as *'sheeps'*, or say *'mans'* instead of *'men'*. They also use tenses in the correct grammatical form, but with the wrong words, such as *'runned'* instead of *'ran,* and *'goed'* instead of *'went'*, or *'gooder '* instead of *'better.'* They are most unlikely to have heard these words from adults, so they are inventing them. Common errors like these have led to the conclusion that babies are pre-programmed to construct language.

However, other scientists contradict this idea. Geoffrey Sampson[22] argues that young children are good at learning language simply because they are good at learning. In the nature versus nurture debate, there is also the conviction that learning a language is entirely an environmental phenomenon.

REFERENCES

[1] Yehudi Menuhin Music Guides, *Voice*, Edited by Sir Keith Falkner (2006) new edition, Kahn & Averill, ISBN 10. 1871082544

[2] Nettle, D., (2004) in *Learning and Language*, SD226 Biological Psychology, Chapter 2 From Sound to Meaning: Hearing, Speech and Language, The Open University, pp. 50-57

[3] Wuethrich, b. (2000). Learning the world's languages – before they vanish. *Science*, **288**, pp. 1156-1159

[4] Carlson, N.R., (1994) *Physiology of Behaviour*, Fifth Edition, Allyn and Bacon, p.512

[5] Peterson, S.E., Fox, P.T., Posner, M.I., Mintun, M., and Raichle, M.E. Positron emission tomographic studies of the processing of single words. *Journal of Cognitive Neuroscience*, 1989, **1**, pp. 153-170, and Price, C., Wise, R., Ramsay, S., Friston, K., Howard, D., Paterson, K., and Frackowiak, R. Regional response differences within the human auditory cortex when listening to words. *Neuroscience Letters*, 1992, **146**, pp. 179-182

[6] Nishimura, H., Hashikawa, K., Doi, K., Iwaki, T., Watanabe, Y., Kusuoka, H., Nishimura, T., and Kubo, T. (1999) Sign language 'heard' in the auditory cortex. *Nature*, **397**, p.116

[7] Pinker, S., (1994) *The Language Instinct*, Penguin Press 1995, p.306

[8] Gardner, R.A., and Gardner, B.T. (1969) Teaching sign language to a chimpanzee. *Science*, **165**, pp. 664-672 and also Gardner & Gardner, and

Premack, D., (1971) Language in a chimpanzee? *Science, 172* pp 808-822, and Savage-Rumbaugh, E.S. (1993). *Language comprehension in ape and child.* Chicago: University of Chicago Press, and Fitch, W.T., and Hauser, M.D. (2004). Computational constraints on syntactic processing in a nonhuman primate. *Science*, **303**, pp. 377-380

[9] Seyfarth, R.M., Cheney, D.L., and Marler, P. (1980) Vervet monkey alarm calls: Semantic communication in a free-ranging primate, *Animal Behaviour,* **28,** pp 1070-94

[10] Tattersall, I., (2002) *The Monkey In The Mirror: Essays On The Science Of What Makes Us Human*, Oxford

[11] http//www.kaitaia.com/jokes/Funny_Lists

[12] McCarthy, R.A., and Warrington, E.K., Evidence for modality-specific meaning systems in the brain. *Nature,* 1988, **334,** pp. 428-435

[13] Kim, K.H., Relkin, N.R., Lee, K. M., and Hirsch, J. (1997). Distinct cortical areas associated with native and second languages. *Nature,* **388,** pp. 171-174

[14] Iverson, J. M., and Goldin-Meadow, S. (1998). Why people gesture when they speak. *Nature,* **396,** p. 228

[15] Rosenzweig, M.R., Breedlove, S.M., Watson, N. V., (2005) *Biological Psychology*, Fourth Edition, Sinauer Associates, Inc., p. 593

[16] Taipale, M., Kaminen, N., Nopola-Hemmi, J,m Haltia, T., et al. (2003) A candidate gene for developmental dyslexia encodes a nuclear tetratricopeptide repeat domain protein dynamically regulated in brain. *Proceedings of the National Academy of Sciences, USA,* **100** pp. 11553-11558

[17] McCarthy, R.A., and Warrington, E.K. (1990) *Cognitive neuropsychology: A clinical introduction.* San Diego, C.A: Academic Press

[18] Rosenzweig, M. R., et. al (2005) op.cit. p.595-6

[19] Fisher, S.E., et. al. Nat. Genet 18 (1998) pp. 168-170

[20] Lai, C.S.L., Fisher, S.E., Hurst, J.A., Vargha-Khadem, F., and Monaco, A.P. (2001) A forkhead-domain gene is mutated in a severe speech and language disorder, *Nature*, **413**, pp. 519-23

[21] Pinker, S., (1994) op. cit. p. 263

[22] Sampson, Geoffrey, (1007) *The 'Language Instinct' Debate*, Continuum International

CHAPTER 10

LEARNING and MEMORY

How do we learn?
How do we remember?
Where and how are memories 'stored' in the brain?
Why do we forget?
How do we remember smells?
How can we improve our memory?

How do we learn?

Learning is a process that potentially changes the way we behave. There are various theories how this takes place. Much of our knowledge about the way we learn has come from experiments with animals.

Associative learning has become connected with the scientist Ivan Pavlov following his experiments with salivating dogs, and is known as conditioning. Effectively it is learning to carry out a specific response whenever a certain stimulus is received. In his experiments Pavlov rang a bell whenever he presented a bowl of food to a dog. Under normal circumstances a dog will salivate at the sight of food, but not when hearing a bell. After repeatedly presenting the two together, the dog associates the two stimuli, and will begin to salivate when he hears the bell alone. This is called a conditioned response, and is classical conditioning.

Another form of learning is called instrumental or operant conditioning. Animals are placed in a Skinner's Box (named after the scientist B.F. Skinner) and would learn, after several trial-and-error attempts, to peck (pigeons) or press (rats) at a lever in order to obtain food pellets.

There are also three types of non-associative learning: habituation, perceptual and motor.

Habituation is a process whereby electrical activity in the brain is decreased after repeated stimuli. Effectively, the novelty of the stimulus wears off. This is an efficient method of saving energy. An example of this type of learning is hearing a clock chime every hour. After a while, we stop noticing the chime if we are doing other things. However, a loud noise such as a window banging shut will startle and you will probably be more aware the next time the clock chimes. You have become temporarily dishabituated, as you have learned to be more aware of noises around you. In this case electrical activity in the brain may increase its original signal strength as you have become more alert to the chime because of the additional input from the window. This is called sensitisation.

Perceptual learning helps us to recognise what we have previously experienced, such as people, objects or situations. There are changes in the sensory association cortex of the brain that detect specific stimuli, and enable us to remember sounds, sights and smells that we can associate with an earlier event. We can also recognise new things such as a friend's new hairstyle.

Motor learning occurs when we need to carry out new movements. Here there are changes in the neuronal circuits in the brain controlling the way we behave (see chapter on Movement).

How do we remember?

There are three major categories of memory: working, short-term and long-term. Figure 10.1 shows the amounts of time it takes to forget these memories.

Figure 10.1 Major temporal categories of human memory

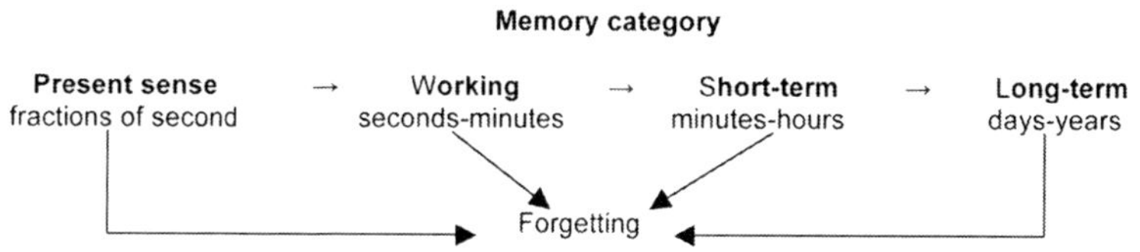

Memory category

Present sense (fractions of second) → Working (seconds-minutes) → Short-term (minutes-hours) → Long-term (days-years)

Forgetting

The sense of the present is a general awareness of surroundings that momentarily slips into consciousness and then disappears, as the detail is not required. For example, you may superficially notice a person asking directions in the street, and then completely forget the event.

Working memory is short-lived and needed for an immediate purpose. For example, you may need to dial a new telephone number. You look it up and then keep repeating it in your head until you have dialled it, and then the number is forgotten. If you need to keep re-dialling the number throughout the day, it will eventually be recalled without having to look it up in the directory. However, if you don't need it again for several days, it will be forgotten. What is happening is that there is a temporary increase in electrical activity within the neuronal networks in the brain, some of which become strengthened, enabling recall. Repetition of the action will strengthen the connections sufficiently for them to be installed within short-term and eventually long-term memories. There are of course many numbers we have redialled many times and can recall at will. This may change with the advent of the mobile 'phone where all numbers are stored in its memory – so we won't need to remember.

Still in the area of numbers, working memory is also able to work out a mathematical problem, such as adding together 39 and 47. You would need to access long-term memory to recognise the digits and remember the method. You may have started with the units 9 and 7 making 16. Your working memory stores the 6 and carries over the 1. Long-term memory then adds the 3 and 4 to make 7 and working memory recalls the carried 1 to make 8 and recalls the 6 to make a total of 86. Or, you may have found it easier to add 40 and 47 and take away the 1. All of this is of course done in a fraction of a minute if you are reasonably competent at mental arithmetic.

Short-term memories are new and have not yet become well established. If there is no further use for them, the connections will decay and be lost. If they are needed again, the memories become consolidated and will remain long-term.

Long-term memory appears to be unlimited in capacity and duration. It is continually being updated and revised as new bits of information alter the information previously stored. There are two main classifications of memory: declarative and non-declarative. Figure 10.2 shows how these are separated.

Figure 10. 2. Long-term memory classification

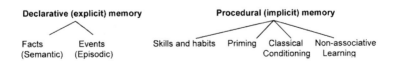

Long-term memory can be divided into two distinct groups: 'knowing that' (declarative) and 'knowing how' (procedural). Declarative memories can be also divided into two sub-types: semantic and episodic[1].

Semantic memories are about knowing facts like the Earth is part of the Solar System and revolves around the Sun. It is encyclopaedic and would cover knowledge of history, geography, science, etc. Episodic memories involve autobiographical events that have occurred in our lives. These may include special dates such as graduation or special birthdays, and events such as family marriages and deaths. They could also relate to a period in hospital or the first day at work. Declarative memory is also known as explicit memory because it relates to particular things.

Procedural memories are implicit. They relate to the skills and habits we acquire, such as learning to read and write or driving a car. They are acquired more slowly and once memorised, may be used automatically. Priming is where exposure to a stimulus enables a response when a similar stimulus is repeated. For example, you are more likely to recall a name or word if you heard it recently. Classical conditioning and non-associative learning are also implicit.

Where and how are memories 'stored' in the brain?

Memories are not 'stored' in the sense that there is one place that holds a particular memory for safe-keeping. Long-term memories change the structure and performance of neuronal circuits in the brain. Each event or stimulus produces a unique pattern of neuronal networks. When reinforced by repetition and learning, these patterns remain constant until needed for recall.

Figure 10.3 is an example of a very simple neuronal network. Each connection, shown by an arrow can either produce more or less neuronal firing.

Figure 10.3 Simplified model of a neural network

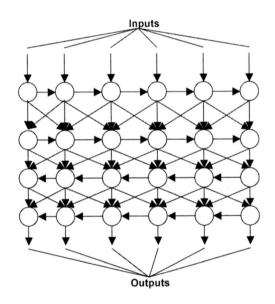

Neural networks are able to discriminate between stimuli by producing different patterns of connections for each. They are also able to trigger the whole reaction when it receives only part of a recognised signal. This is a very energy-efficient method of working. Damage to parts of neuronal connections degrades only that part and leaves the remainder of the network in place.

Between each neuron is a gap known as a synapse or synaptic cleft, and it provides the link between the two. Electrical signals pass along a neuron until they reach the end, and this is called the pre-synaptic neuron. Chemical neurotransmitters are released which interact with receptor proteins on the postsynaptic neuron. The resulting chemical mix between that and many other neurons will either excite the synapse and enable the signal, called an action potential to be passed to the next neuron via electrical synapses, or the signal will be inhibited and prevented from travelling further. The psychologist Donald Hebb[2] suggested that if a post-synaptic neuron fires at the same time as the pre-synaptic terminal releases its transmitter, then the connection would be strengthened, and result in learning and memory retention. This type of connection has been named after him and is called a Hebbian Synapse. Figure 10.4 shows a diagram of chemical and electrical transmission at a synapse.

Figure 10.4 Chemical and electrical synaptic transmission

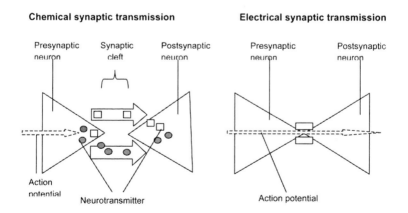

There are various areas in the brain that are involved in memory. Figure 10.5 shows some of these. Patients undergoing surgery recall a specific memory when a part in the back of the temporal cortex is electrically stimulated. This suggests that this area is concerned with certain memories. Experiments with non-human primates show the cortex is involved with vision and touch processing as well as memory location. Damage to part of the temporal cortex disrupts memories associated with vision, leaving other types of memory intact.[3]

Long-term potentiation (LTP) is a change in the reactivity of a postsynaptic neuron that lasts for hours or days and results in a long-term memory. There are five stages that lead to the production of LTP, and they are collectively called a neurochemical cascade:

Stage 1. Strong stimulation of the neuron leads to an increase in concentration of calcium ions (Ca^{2+} ions) at the synapse. An ion is a molecule that has acquired an electrical charge by losing or gaining electrons (minute particles charged with electricity).

Stage 2. The increased calcium ion concentration activates a class of enzymes called kinases that adds phosphorous groups to proteins.

Stage 3. These phosphorated CREB proteins bind to parts of genes and carry information by transcription.

Stage 4. The structural protein changes strengthen the synapses in the brain, and induce LTP.

Stage 5. Some of these synthesised proteins are transported

along the nerve branches, which can alter the responses of other stimuli.

Figure 10.5 Areas of the brain involved in learning and memory

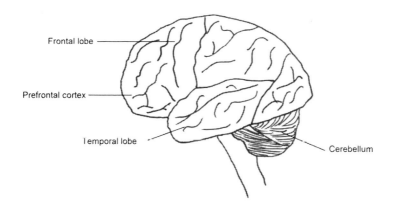

If the middle of the temporal lobe is damaged, there is an inability to assimilate new declarative information, either episodic or semantic.[4] This is particularly noticeable when the neuronal connections between the frontal and temporal lobes, and the hippocampus and mammillary bodies are disrupted. The mamillary bodies are a pair of nuclei found at the base of the brain.

The prefrontal cortex is involved in the management of memory. Memory searches are guided and actions planned. The prefrontal cortex also plays a part in distinguishing between true and false memories, and damage to this area can create false memories[5].

The cerebellum is a structure at the back of the brain and damage to this also affects procedural learning. Positron Emission Topography (PET) scans have shown that during this type of learning there is an increased blood flow to the cerebellum.[6]

The Limbic System is a network of structures involved in emotion and learning, including the hippocampus, the amygdala and the mamillary bodies (see also chapter on Emotions and Stress). Figure 10.6 shows the location of the organs involved in learning and memory. The hippocampus is a part of the cerebral hemispheres curled in the base of the middle of the temporal lobe. It receives information from and projects to various cortical areas, and is involved in establishing memories at the various cortical areas. People with damage to this area are able to accurately draw an object shown in a mirror (reverse mirror drawing), but when asked to do it again, cannot remember doing it before.

Figure 10.6 Part of the Limbic system and the Basal Ganglia

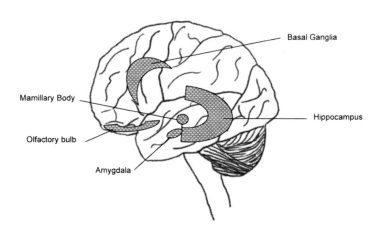

Interesting experiments using imaging technique scans have been carried out on London Taxi Drivers.[7] Part of their training involves memorising all the streets of London. The overall size of the hippocampus remained normal, but the longer a person had been driving a taxi, the larger the back part of the hippocampus, especially the right side became, compared to non-taxi drivers. The front part of the hippocampus however was smaller than the non-taxi drivers. The posterior part of the hippocampus is involved in spatial memory, and the experienced driver accumulates enlarged neuronal circuits to accommodate the additional knowledge needed to carry out the job.

The amygdala is a group of nuclei situated in the temporal lobe, which is an important site relating to emotion and memory. An event which frightens or thrills us is much more likely to be remembered. Noradrenaline is a synaptic neurotransmitter, and increased amounts are secreted when memories become established.

People with post-traumatic stress disorder appear to be caught up in a positive feedback loop that re-lives the experience, producing hormone stress responses that reinforce the memory[8]. Symptoms include nightmares and flashbacks. Neuro-imaging techniques have revealed the disorder is associated with a hyper-arousal of the amygdala, which is insufficiently controlled by processes in the medial frontal cortex and the hippocampus.

The basal ganglia is another group of nuclei situated deep inside the cerebral cortex. Damage to this area can result in disruption of procedural memory, but declarative memories remain intact. Patients are unable to learn new skills and other non-associative learning.

If the brain stored all our memories, it would be unnecessarily overloaded. Therefore there is a benefit in realising what to retain and what to forget. A study was carried out on a memory expert (mnemonist), who could recall lists of over seventy words or numbers with ease and over a period of several days[9]. He explained that he produced an internal visual image of each item, which aided his memory. Voices or musical sounds could become colours or tastes. He also used the strategy of memorising a familiar location and placing each item somewhere in a familiar street or room. He was also able to remember subjects that were of no meaning to him, such as poems in an unfamiliar language. However there was a downside to his remarkable memory. He was unable to do what people with normal memories can do with ease. He had difficulty in reading, because the words automatically created complex images in his brain. He was unable to group items into categories. For example, he could not recognise the alphabet as a list of all the letters used in a language. To him it would be a meaningless list of letters, each of which needed a special image to be remembered. He was also unable to separate trivial pieces of information from those of importance, and would sometimes spend an inordinate length of time over such irrelevant items.

Why do we forget?

It seems incredible that we can often recall the words of a song or person's name many years since we first heard it. A smell can trigger a memory from our distant past. We could be forgiven for believing that everything we ever learned remains somewhere in our brain, just waiting to jump back into consciousness at the right moment. But it is also very obvious, especially as we get older that our memories often let us down. We cannot remember a word or certain event in the past. Psychologists have suggested several theories that could account for our inability to recall at will, such as the Interference Theory, the Repression Theory, the Decay Theory, the Retrieval Failure Theory and the Consolidation Theory.

The Interference Theory. Proactive interference is where material we have learned interferes with a newer memory, and retroactive interference is where a new memory interferes with something we learned earlier. Forgetting can be caused by a decline of the long-term potentiation signals at the synapses caused by these types of interference.

The Repression Theory was proposed by the psychoanalyst Sigmund Freud. He argued that we bury memories that are threatening or make us anxious. He believed these memories could be fully recovered through dream interpretation, by hypnosis or by free association of ideas at psychoanalysis sessions. However, recent research indicates that it is unlikely we can completely regain these memories, as the stressful circumstances surrounding the event will result in poor memory. There is also a distinct possibility that the recalled memories are in fact false, or distorted.

The Decay Theory suggests that forgetting is caused by the passing of time. If information is not used, the memory trace will decay. However this has not been substantiated, as it does not account for the memories that can easily be recalled many years later, without being used in the meantime, as mentioned in the opening paragraph of this section.

The Retrieval Failure Theory specifies that the information in memory cannot be accessed because of inadequate retrieval cues. The more similar the cues the easier it is to recall the memory. There is also evidence that mood-states affect memory in the absence of other retrieval cues.[10]

The Consolidation Theory assumes that recently formed memories are still being consolidated and during this period are more susceptible to interference and forgetting. Some evidence of this is that people who drink alcohol to excess are likely to suffer from memory loss, although they can recall events prior to the drinking session. This theory explains why the rate of forgetting decreases over time, as it is the most recent memories that are most vulnerable to loss.[11] However, there are contradictions to this theory as it relates to amnesia.

Retrograde amnesia is the inability to recall what happened for a period before a traumatic event, sometimes stretching back over several years. These memories would have been held in long-term memory, and should therefore not be too fragile to be vulnerable to interference. Some memories reappear, indicating there is a problem with retrieval rather than memory storage. Also there are pockets of memory within the time-frame of the amnesia, which are available for recall, so there is not a completely blank slate in that period. Anterograde amnesia is a failure to form new memories after a traumatic event.

Amnesia can result from an injury to the brain areas involved with episodic and declarative memories, including the medial temporal lobes and the hippocampus. Long-term alcoholism, as mentioned above, can also promote retrograde amnesia. Malnutrition, especially a deficiency of vitamin B1 is another cause of amnesia.

A report in 2010 concluded there are 820,000 people suffering from Alzheimer's disease in the U.K[12]. It affects about 10% of the population over 65 and up to 45% over 85.[13] Symptoms are impairment of recent declarative memory and an inability to concentrate. The condition is progressive and will adversely affect language skills, visual and spatial orientation and abstract thinking.

Post-mortem examination shows there is a loss of neurons, especially in the brain regions associated with learning and memory. There are also large deposits of an abnormal protein called beta-amyloid, sometimes called amyloid or senile placques, which occur outside the cells. Beta-amyloid can also be a genetic factor. Within the cells are abnormal groups of a protein called tau, which form neurofibrillary tangles. Both the placques and the tangles are toxic and will eventually kill neurons.

It has been thought that the ageing process itself can impair learning and memory functions. However, this is difficult to measure correctly, as other factors may play a part. These may be educational levels, motivation, or poor health. Research has shown that an enriched environment throughout life can slow-down or prevent the decline of cognitive faculties in old age. A comprehensive study of more than 5,000 people over a 35-year period suggests that having complex and interesting lifestyles, strong family ties and no financial concerns all contribute to maintaining a strong and intact memory and learning ability.[14] However, the 2010 report mentioned above indicates that diet and exercise may play a more important role in combating the disease than brain stimulation.

How do we remember smells?

In 2006 a team of scientists in the United States claimed to have found clues to the way the brain remembers smells. The study indicated that special cells called Blanes cells could retain information about a smell, and magnify the brain's response by triggering electrical firing in the neurons. The individual cells then retain the memory of the smell by storing the information.[15]

Molecules of an odour enter the nose and interact with sensory neurons, which send a signal to the olfactory bulb (see Figure 10.6). This is the part of the brain associated with smell. The team found in their study using the brains of rats, that the olfactory bulb received magnified signals from the patterns of connections. The way this part of the brain works is very similar to the cortical region of the brain associated with memories, and is also the same part of the brain that is damaged in Alzheimer's disease. It may explain how smells can often trigger memories. However, no studies have yet been carried out on humans, so the effects could be different.

How can we improve our memory?

There are various strategies for enhancing memory, some of which were used by the mnemonist discussed above. Items on a list are placed in our thoughts into a location, such as the rooms of a house, or a well-known route. When trying to remember, we think about walking around the house or route noting each item as it is passed. Using a rhyming image for each item is another method, such as one is the sun, two is a shoe, three is a tree etc. For example, if the first word is smile, the image is a smiling sun, if the second word is buckle it is imagined stuck onto a shoe and so on. Another technique is linking words into a story. As the story unfolds, the words are recalled. Another routine is using initial letters to remember a sequence. An example would be remembering the colours of a rainbow by using the sentence **R**ichard **O**f **Y**ork **G**ains **B**attles **I**n **V**ain for the colours Red, Orange, Yellow, Green, Blue, Indigo and Violet. Finally there is the PQRS scheme much favoured by all those students before taking exams. Preview, Question, Read and Summarise.

REFERENCES

[1] Tulving, E., (1972), Episodic and semantic memory. In E. Tulving and W. Donaldson (Eds), *Organization and memory*, Academic Press, New York, pp. 381-403

[2] Hebb, D.O. (1949), *The organization of behaviour*, Wiley, New York

[3] Petrides, M., (1994), Frontal lobes and working memory: evidence from investigations of the effects of cortical excisions in nonhuman primates. In *Handbook of Neuropsychology*, **Vol. 9** (ed. F. Boller and J. Grafman), Elsivier, Amsterdam, pp. 59-82

[4] Milner, b. (1966), Amnesia following operation on the temporal lobes. In *Amnesia* (eds C.W.M. Whitty and O.L. Zangwill) Butterworths, London pp. 109-133

[5] Schacter, D.L., (1997a), The cognitive neuroscience of memory: perspectives from neuroimaging research. *Philosophical Transactions of the Royal Society of London B*, **352**, 1689-1695 and (1997b) False recognition and the brain. *Current Directions in Psychological Science*, **6**, 65-69

[6] Grafton, S.T., Woods, R.P., and Tyszka, M., (1994), Functional imaging of procedural motor learning: relating cerebral blood flow with individual subject performance. *Human Brain Mapping*, **1**, 221-234

[7] Maguire, E.A., Gadian, D.G., Johnsrude, I.S., Good, C.D., et al (2000), Navigation-related structural change in the hippocampi of taxi drivers. *Proceedings of the National Academy of Sciences*, USA, **97**, 4398-4403

[8] Pitman, R.K., (1989), Post-traumatic stress disorder, hormones and memory. *Biological Psychiatry*, **26**, pp. 221-223

[9] Luria, A.R., (1987), *The mind of a mnemonist*, Cambridge, M.A: Harvard University Press

[10] Kenealy, P. M., (1997), Mood-state dependent retrieval: the effects of induced mood on memory reconsidered. *Quarterly Journal of Experimental Psychology*, **50A**, pp 290-317

[11] Wixted, J.T., (2004), The psychology and neuroscience of forgetting. *Annual Review of Psychology*, **55**, pp. 235-269

[12] www.alzheimers-research.org.uk.The Alzheimers Research Trust Report Univ. of Oxford

[13] Murphy, K.., and Naish, P., (2004), *Learning and Language*, Bk 5., Biological Psychology: Exploring the Brain, Open University, p.24

[14] Schaie, K.W., (1994), The course of adult intellectual development. *American Psychologist*, **49**, pp. 304-313

[15] Pressler, R.T., and Strowbridge, B.W., (2006), Blanes Cells Mediate Persistent Feedforward Inhibition onto Granule Cells in the Olfactory Bulb., *Neuron*, **49**, pp. 889-904

CHAPTER 11

MOVEMENT

How do we move?
Which are the main bones of the body?
What does bone consist of?
Disorders affecting bones
What are the different types of joint?
What makes a joint crack?
Problems affecting joints
What are the different types of muscle?
How does muscle action work?
What makes a muscle tired?
What causes the knee-jerk response?
How does the brain control movement?
What causes Parkinson's and Huntington's diseases?
Other disorders affecting movement

How do we move?

There are basically two kinds of movement: autonomic, which is carried out without conscious awareness, and voluntary. Both kinds are made in response to signals received from sensory and motor neurons involving the brain and spinal cord. To carry out a movement requires the co-ordinated effort of bones, joints and muscles.

Which are the main bones of the body?

The human skeleton consists of a trunk, including the head, neck, chest (thorax), shoulder girdle and upper limbs, pelvic girdle and lower limbs. Its main functions are to provide a support and shape for the skin and muscles and to protect the internal organs. Figure 11.1 shows the bones making up the skeleton. Human babies are born with over 300 bones. Some of these fuse during development, such as the sacrum and coccyx at the lower end of the spine and the skull. Adults have 206 bones.

The skull consists of the cranium, which has 8 flat fused bones housing the brain, and the face, which has 14 bones forming the cheeks, upper and lower jaws, various sides of the nose, the roof of the mouth and the eye sockets.

Figure 11.1 Human Skeleton

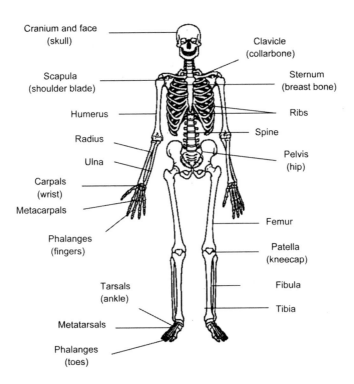

Figure 11.2 shows that the spinal column consists of individual vertebrae. The first bone is called the atlas vertebra and supports the skull. The second bone is called the axis vertebra and allows the head to rotate. These bones have synovial joints that are held together by ligaments. Synovial fluid between the bones helps the joints to move effortlessly. The remaining vertebrae have joints made up of cartilage pads that allow a very limited degree of movement. The 12 thoracic bones have 12 pairs of ribs attached. The first 7 are attached to the sternum (breastplate) and are called true ribs. The next 3

pairs of ribs are attached to the ribs above them and are called false ribs. The last 2 pairs of ribs are only attached at the back and are called floating ribs.

The hands and feet are each divided into three main regions that are similarly arranged. 8 small carpal bones make up the wrist, whilst there are 7 tarsal bones making up the ankle. The next group, the metacarpals in the hand and the metatarsals in the foot forming the body of the hands and feet, consist of 5 bones for each limb. Finally there are 14 phalange bones in each limb that make up jointed fingers and toes. Each contains 3 small bones, except the thumbs and big toes, which have two.

Figure 11.2 The Spinal Column

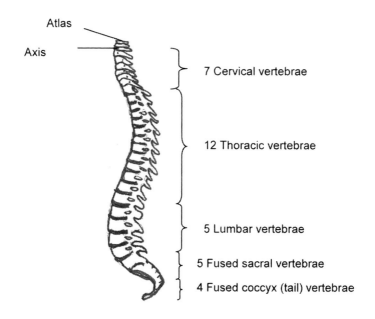

What does bone consist of?

Bones are made of connective tissue consisting of calcium salts and calcium phosphate, and organic osteoid that is mainly collagen. There is also a supply of blood and nerves. There are five different types of bone: long, short, irregular, flat and sesamoid. Long bones are those in the arms and legs. Short bones are found at the wrist and ankle. Spinal vertebrae and the pelvis are examples of irregular bone. Flat bones are the sternum (breast bone), the ribs and the fused bones at the top of the skull. Sesamoid bones such as the kneecap and the small knobbly bones at the wrist are embedded in tendons.

Bones grow throughout childhood and when mature ossify (harden) into hard structures.

There are two types of bone tissue: compact and cancellous (spongy). Compact bone tissue is hard and durable and is made up of tubular units called osteons. Long bones have a shaft with hard compact bone tissue. There is a central hollow canal containing yellow marrow, which is a storage area for fat. Cancellous bone tissue has a honeycomb appearance, and the spaces are filled with red bone marrow, and produces new blood cells.

Under exceptional circumstances, after severe loss of blood, the body can convert yellow fatty marrow into red marrow to increase the production of new blood cells.

All bones have both types of tissue in various combinations, and are covered by a membrane of connective tissue called the periosteum. It has a fibrous outer layer with a rich blood supply and an inner layer that produces new cells for growth and repair of bone tissue.

There are three types of bone cells. Osteoclast cells break down and assimilate old bone cells at the surface of bones to maintain the best possible shape. Osteoblasts secrete collagen and are found in the deeper layers of the periosteum, and also where the bone has fractured. They form new bone cells. Collagen is a protein that strengthens blood vessels. As the cells mature, they develop into osteocytes that prevent the formation of new bone, by monitoring and maintaining bone tissue.

Disorders affecting bones

Osteoporosis is a condition whereby bone density is reduced. New bone is not developed quickly enough to replace bone cells that have been absorbed, resulting in a loss of bone structure. It is more prevalent amongst the elderly and especially post-menopausal women. There is an increased risk of bone fracture and loss of height with age, as the vertebrae become more compressed.

Paget's disease is caused by abnormal activity of the osteoblasts and osteoclasts. The bone breaks down more quickly and becomes softer than normal bone. The affected bones bend or fracture more easily. This usually occurs after the age of 40 and affects the skull, pelvis and leg bones.

Rickets in children and osteomalacia in adults is caused by a deficiency of vitamin D. It results in deformity of lower limbs. Rickets may be caused by poor diet, lack of sunlight or a genetic predisposition that prevents vitamin D from being metabolised.

Osteomyelitis is a bacterial bone infection, which can occur after a surgical operation or after a bone fracture. It is most commonly found in children.

What are the different types of joint?

A joint is the junction between two bones that allows movement to occur. There are three main types of joint: fibrous, cartilaginous and synovial

 Fibrous joints are linked with a tough connective tissue. In the skull the sutures allow no movement in adults (although there is a very limited amount of movement in the very young to allow for growth).

Cartilaginous joints have a pad of fibrocartilage that acts as a shock absorber. They are slightly moveable as in the vertebrae of the spine.

Most of the joints in the body are synovial. The cavity between the bones contains synovial fluid which lubricates the bones to allow for ease of movement. The cartilage tissue at the ends of the bones, are held together by ligaments to help provide stability. Figure 11.3 shows some of the locations of the different types of synovial joint.

There are six types of synovial joint:

Hinge joints are similar to a hinged door opening and closing.

The convex surface of one bone fits into the concave surface of the adjacent bone. The knee and elbow are examples of this type of joint.

Figure 11.3 Skeleton showing location of synovial joints

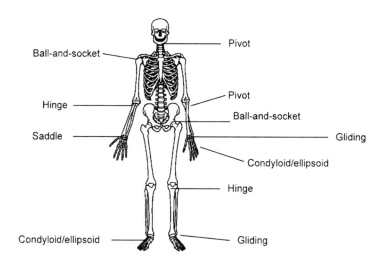

Gliding joints are flat or slightly curved and held together by ligaments. They glide over each other, allowing a limited degree of movement. The small bones of the wrist and ankle, and the vertebrae of the spine are examples of this type of joint.

Pivot joints allow a limited circular rotation. The rounded surface of one bone rotates within a ring of a ligament. The radius and ulna of the lower arm and the atlas and axis (first and second) cervical bones in the neck are examples of this type of joint.

Ball-and-socket joints allow movement in all directions. The ball-shaped end of one bone fits into the cup-shaped adjacent bone. The shoulder and hip are examples of this type of joint.

Saddle joints sit together like a saddle on a horse. They allow movement in all directions. The base of the thumb where the carpal meets the metacarpal is an example of this type of joint.

Condyloid or **ellipsoid** joints sit in a cup-like depression on the adjacent bone. They allow bending, extending and some side-to-side rotation. The metacarpal to phalange joints in the hand and the mandible to temporal bone of the jaw are examples of this type of joint.

What makes a joint crack?

Some joints such as knees or knuckles make a cracking noise when extended. This is caused when a bubble formed in the synovial fluid bursts due to a change in pressure, created by the movement.

Problems affecting joints

Rheumatoid arthritis is a progressive inflammatory disease that mainly affects peripheral synovial joints. It is an autoimmune condition whereby the immune system attacks the joints usually of the hands and feet. It is more prevalent in females and there is a strong genetic connection, as it is often found in several members of the same family. Sufferers have high levels of an antibody called rheumatoid factor (RH).

Osteoarthritis is a degenerative disease with little or no inflammation. It is usually found in those aged over 65, and is caused when cartilage becomes worn down. It also has a heredity factor and is exacerbated by repetitive use of affected joints. Women are again more likely to contract the condition. Obesity is another high risk feature.

Gout occurs mainly in males and also runs in families. It is caused by deposits of sodium urate crystals in joints and tendons, and produces inflammation. Obesity and high alcohol intake are factors known to increase the likelihood of the condition occurring.

What are the different types of muscle?

There are three types of muscle tissue: cardiac, smooth and skeletal. Cardiac muscle is only found in the heart. It has an involuntary action causing the heart to expand and contract, pumping blood through blood vessels and creating the heartbeats.

Smooth muscle also has an involuntary action. Its cells are small and spindle-shaped with a single nucleus. It is found in the walls of hollow organs and changes their diameter by an action called peristalsis. Smooth muscle is responsible for pushing food through the digestive system and eliminating waste material.

Skeletal muscle is under voluntary control. It is attached to tendons and is used to move the bones of the skeleton. It has two types of muscle fibre producing a striped appearance under the microscope. Fast twitch or white muscle fibre contracts quickly for short periods, and uses glycogen for power. It provides strength and speed, but tires quickly. This type of muscle is found in the eye. Slow twitch or red muscle fibre contracts slowly for sustained periods, and is thinner than fast twitch. It uses oxygen for energy and is found in the muscles of the legs, trunk, back and hips. These muscles provide stability to maintain posture. Most muscles contain a mixture of both types. The amounts of each vary genetically. Long-distance runners will have more slow twitch fibres, whereas sprinters have a high proportion of fast twitch fibre allowing them to run short distances fast [1]

There are four kinds of skeletal muscle: agonist, antagonist, synergist and fixator. Agonists create the contraction necessary to move and antagonists oppose the action by relaxing the muscles. They lie on opposing sides of the bone. So for example to flex the arm, the agonist biceps flexor muscles contract, and the antagonist triceps muscles extend. Synergist muscles are small and assist the prime agonists to control the level of movement to make it appropriate for the task. Fixator muscles are large and prevent the body from toppling over following a movement. They help the body stabilise and maintain its balance.

How does muscle action work?

Skeletal muscles consist of thousands of fibres with two kinds of filaments that are arranged along the length, giving them a striped appearance. The filaments comprise of proteins called myosin and actin. The myosin produces a thick layer alternating with a thin layer of actin. Repeating units of thick and thin layers are called sarcomeres. At each side of the sarcomere is a stripe called a Z-line. Figures 11.5a) and 11.5b) shows the structure of a sarcomere, relaxed and contracted. When a muscle contracts, the filaments slide past each other, shortening its length. Each muscle fibre is a cell containing many nuclei. Little organelles called mitochondria inside the cell produce adenosine triphosphate (ATP), which stores and transports energy. There is also a substance called myoglobin that accumulates oxygen inside the muscle. Stores of calcium are also released into the cellular fluid as required for use by the muscle.

Figure 11. 5 a) Sarcomere (relaxed)

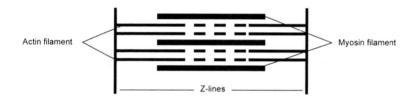

Figure 11.5 b) Sarcomere (contracted)

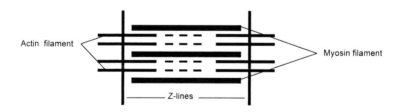

When skeletal muscle cells contract, action potential signals from the motor neurons fire through the sarcomeres, and penetrate into the muscle cell. This releases calcium from the stores inside the cell, and triggers the binding of myosin to actin to form a bridge. ATP provides the energy for the two filaments to slide over each other, pulling the Z-lines closer and shortening the sarcomere.

A muscle spindle is a receptor that lies parallel to a muscle and sends impulses to the central nervous system when the muscle is stretched. Within the muscle spindle are intrafusal muscle fibres that are sensitive to the degree of stretch applied to the muscle fibre. Extrafusal muscle fibres lie outside the muscle spindles and provide most of the force needed for muscle contractions. Stretch receptors are also found in the Golgi tendon organ at the ends of the muscles. They detect how much stretch is needed by the muscle through the tendons attached to the bones.

Figure 11.6 Skeletal muscle

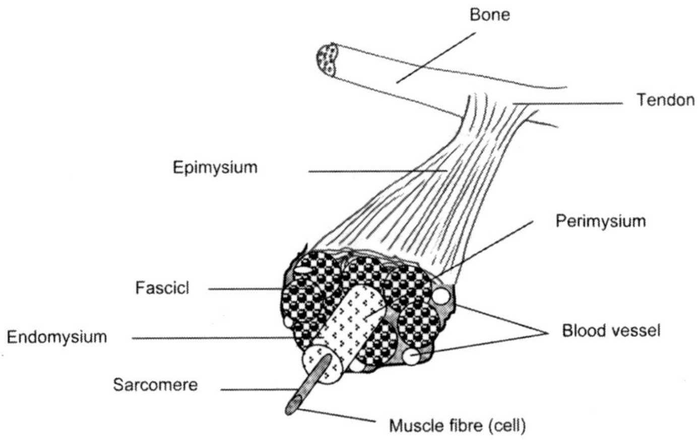

Figure 11.6 shows the muscle structure. Skeletal muscle fibres are covered in a connective tissue sheath called epimysium. The cells inside the muscle are in separate bundles called fascicles. These are covered in connective tissue called perimysium. Individual muscle cells are wrapped in a connective tissue layer called endomysium.

The total mass of muscle fibre remains stable throughout adulthood. However, the muscles can be strengthened and made more flexible through exercise. Muscles that are rarely used tend to waste and are not able to carry out normal levels of activity.

What makes a muscle tired?

Muscles can only work if they have a plentiful supply of oxygen and other fuel such as glycogen produced from glucose. When they are well exercised, the fuel levels become too low for the muscles to function properly. They become fatigued and start to ache. Lactic acid is formed and accumulates in the muscle tissue. It also causes breathlessness. As muscle action slows down, deep breathing enables the intake of more oxygen to supply the muscles. The excess lactic acid is removed through the blood vessels to be broken down by the liver, which stores it as glycogen. This will be used as needed.

What causes the knee-jerk response?

The knee-jerk stretch reflex is an involuntary action, which occurs when the knee-cap is tapped. The tendon runs from the thigh, over the patella (knee-cap) and down the lower leg. The quadriceps muscle in the thigh contracts, making the lower leg move forward.

When the tendon is stretched, the muscle spindle receptors trigger impulses in the sensory neuron of the femur (upper leg) leading to the lumbar region of the spinal cord and synapses with a motor neuron via a relay neuron. Impulses are then sent back to the muscle in the femur, which starts the contraction. This in turn co-ordinates with the relaxation of the antagonistic hamstring tendon of the lower leg, causing the leg to kick forward.

Figure 11.7 Knee-jerk reflex

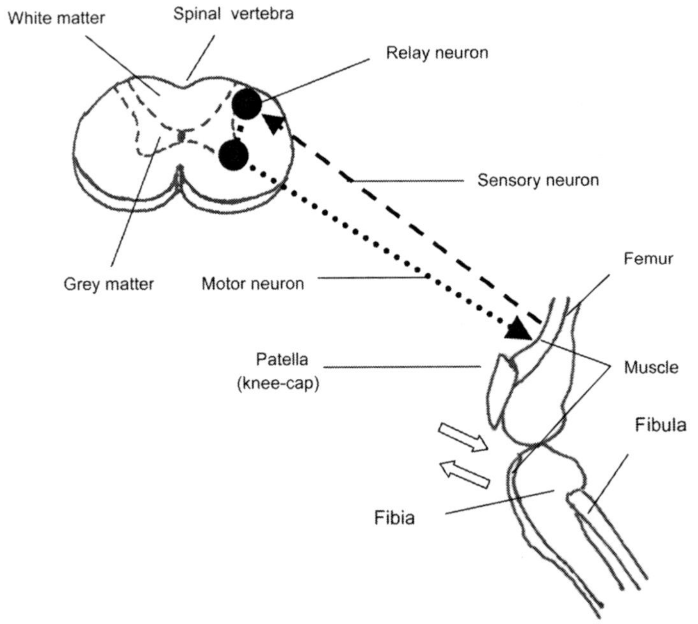

How does the brain control movement?

The brain and the spinal cord have different motor pathways that can allow us to make several movements simultaneously. For example we can walk and talk and wave our hands about at the same time. The primary motor cortex, in association with the pre-motor areas and supplementary motor areas of the brain plan and carry out actions. The somatosensory cortex processes information related to touch. Figure 11.8 shows the approximate positions of these regions. Large neurons called Betz cells send signals along long axons down the spinal cord to synapse with motor neurons which connect to muscles. Various reflex circuits link areas of the body along different sections of the spinal cord.

Figure 11.8 Cortical areas involved in movement

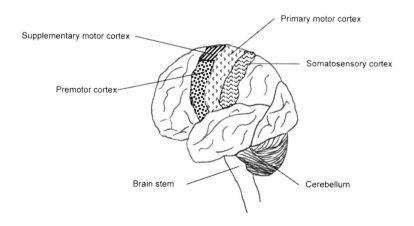

The pre-motor and basal ganglia areas are found in the primary motor cortex. They are involved not only in planning actions but also refining movements. These are based on sensory information received from the cerebellum and the brain stem, via the thalamus. One theory is that the basal ganglia selects the appropriate action to be carried out.[2]

The cerebellum integrates the sensory-motor systems, and some pathways may be involved in planning potential movements.

The brain stem is involved in the control of posture.

The primary motor cortex has sections devoted to processing information from various parts of the body. These sections are larger where they control particularly touch-sensitive regions, such as the lips and the fingers. Figure 11.9 shows how the sections relate to different areas of the body.

Figure 11.9. Primary motor cortex

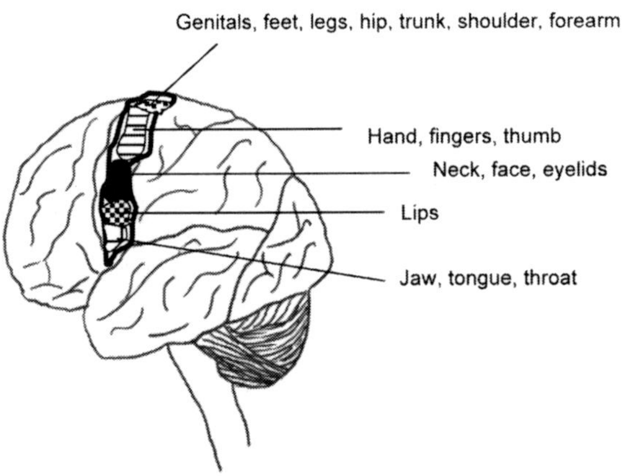

The weight of an object determines the level of force required by the muscles to pick it up. This is controlled by the sensory neurons in the skeletal muscle which travel to the spinal cord. The information is passed to the brain which analyses the signals received and sends back impulses through the motor neurons to activate the muscles. At all times, the body must retain its balance by redistributing weight. If we accidentally move in an inappropriate way, due to a sudden unexpected action, we tend to fall over. Even then we will automatically put out our hands first to protect the the ribs and vital organs in the chest.

The brain learns to adjust the level of force required based on previous experience. When we are learning new movement skills, such as typing or playing tennis, our actions are slow, and awkward. As we become more skilled, they become faster and more routine and spontaneous. During the learning period, the supplementary motor area is highly activated, and becomes less so as we extend our expertise.[3]

When the brain issues instructions to move a limb, the larger alpha motor neurons begin the process of contracting the extrafusal muscle fibres. At the same time, gamma motor neurons innervate the intrafusal muscle fibres controlling the length of the muscle spindles, and the degree of contraction.

What causes Parkinson's and Huntington's diseases?

The symptoms of Parkinson's disease are tremors of the hands and face and muscular rigidity. It is caused by a degeneration of some cells in the substantia nigra area of the brain, which normally release the neurotransmitter dopamine. Dopamine transmits signals to the basal ganglia, which in turn sends signals to the primary motor cortex to trigger movement. The basal ganglia, therefore receives insufficient information to trigger action. [4]

Dopamine cells are too large to cross the blood-brain barrier and enter the nervous system. Dopamine cannot therefore be taken by patients to alleviate their symptoms.[5] The disease is not generally considered to have a genetic basis even though several members of the same family can be affected.[6] It may therefore be due to environmental factors.

The symptoms of Huntington's disease are excessive, uncontrolled, jerky movements of the body. The condition is caused by a dominant gene that is inherited, and leads to degeneration of GABA-ergic neurons and parts of the basal ganglia. These neurons would normally inhibit and manage muscular activity. The disease only affects adults, as it develops slowly. Victims of the condition will pass on the defective gene to 50% of their children.

Other disorders affecting movement

Cramp is caused by an involuntary contraction of the muscles, usually in the leg or abdomen. The causes are mostly unknown. It occurs mainly at night and may be because we tend to sleep with our knees bent and feet pointed down which aggravates the calf muscles. Stretching or moving the legs often overcomes the problem. It is more common in people over the age of 60. Cramp can also be caused by a lack of water or salt. Over-exercising in very cold weather or during pregnancy can create the condition Excess alcohol intake can produce cramp, and some medications can cause it as a side effect.

Motorneuron disease causes the muscles to waste, when neurons in the brainstem and spinal cord degenerate. The condition affects speech, breathing, swallowing and all limb movements. The cause has not yet been established.

Muscular dystrophy causes muscles to waste away. A defective gene is carried on the X chromosome. It mainly affects men, as women carry two X chromosomes, one of which is usually normal, and means they are much less likely to develop the condition.

Myasthenia gravis is a disease of the immune system that weakens skeletal muscles Antibodies block receptors at the neuromuscular junctions, and affects the head and eye muscles and in the later stages disrupts swallowing and respiration.

Restless leg syndrome occurs during periods of inactivity, and the symptoms are an overwhelming urge to move the legs. More common in later life, its causes are not known. There may be a genetic predisposition to the condition, which can be triggered by smoking or by drinking caffeine or alcohol. It may also be related to low iron levels or neurological disorders. Some medications are known to aggravate the condition, such as anti-nausea or anti-seizure drugs.

Snake bites can be venomous and can also block the neuromuscular junctions. Toxins inhibit the enzyme cholinesterase, which is used in muscle control. They also contain enzymes that disrupt the production of the nucleotide ATP used to transfer energy in cells. The effect from the snake's point of view is to make the prey unable to use its muscles to fight back.

REFERENCES

[1] Parson, T., (2005) An Holistic Guide to Anatomy & Physiology, Thomson Learning, p. 92

[2] Graybiel, A.M., and Kimura, M., (1995) Adaptive neural networks in the basal ganglia. In *Models of Information Processing in the Basal Ganglia* (eds J.C.Houk, J.L. Davis and D.G. Beiser), The MIT Press, Cambridge, pp. 103-116

[3] Jenkins, I.H., Brooks, D.J., Nixon, P.D., Frackowiak, R.S.J., and Passingham, R.E., (1994). Motor sequence learning. A study with positron emission tomography. *Journal of Neuroscience,* **14,** pp 3775-3790.

[4] Robbins, T.W., and Everitt, B.J., (1992) Functions of dopamine in the dorsal and ventral striatum. *Seminars in the Neurosciences,* **4,** pp. 119-127

[5] Toates F., (2001) *Biological Psychology,* Pearson Education Ltd., p. 268

[6] Carlson, N.R., (1994), Physiology of Behaviour, Fifth Edn., Allyn and Bacon (Parmount Publishing), p. 342

CHAPTER 12

NERVOUS SYSTEM

What is the nervous system?
What does the central nervous system consist of?
What does the peripheral nervous system consist of?
How do nerve cells work?
What causes seizures?
What causes the sensation of 'pins-and-needles'?
Disorders affecting the nervous system

What is the nervous system?

The nervous system consists of two major parts: the central nervous system that is the processing centre and the peripheral nervous system. The peripheral nervous system is sub-divided into the somatic and autonomic systems, and attends mainly to the limbs and organs.

What does the central nervous system consist of?

The central nervous system consists of the brain and the spinal cord. These structures are protected by bone and three layers of tissue called the meninges. The outer layer is made up of fibrous tissue containing blood vessels and nerves. It covers the outer part of the brain and surrounds the spinal cord. The middle layer forms a space containing cerebrospinal fluid, which nourishes the system. The inner layer contains a rich blood supply for further sustenance. The spinal cord transmits sensory information received by the body from the environment to the brain. It conveys instructions to take certain actions received from the brain to the rest of the body. Information is transmitted via different types of nerve cells.

What does the peripheral nervous system consist of?

The peripheral nervous system transmits information to and from the central nervous system along nerve pathways, and consists of two parts: the somatic and the autonomic systems. The somatic nervous system controls voluntary body movements through the action of skeletal muscles. Afferent nerves lead to the central nervous system which receives sensory information from external sources such as touch, hearing and sight. Efferent nerves carry information away from the central nervous system and are responsible for muscle control.

Twelve pairs of cranial nerves radiate from the brain to all areas of the face and neck. As can be seen from Figure 12.1 each pair of nerves is associated with a different part of the head and body.

Figure 12.2 shows the thirty-one pairs of nerves connected to the spinal cord, each part is responsible for one side of the body. Each nerve has two roots. The front root has pathways that connect the spinal cord to skeletal muscles. The back root connects sensory pathways between the body and the spinal cord.

Figure 12.1 Underside view of brain showing cranial nerves

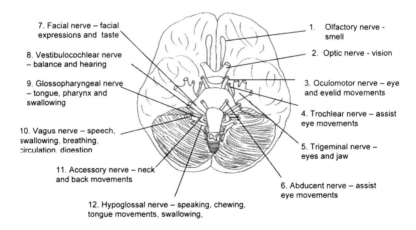

The thoracic nerves supply the muscles of the chest and abdomen. The other groups of nerves form networks called plexuses, which branch out to different parts of the body. The cervical plexus supplies the skin and neck muscles and shoulder. The brachial plexus supplies the skin and muscles of the arms. The lumbar plexus supplies the skin and muscles of the lower abdomen, groin and lower legs. The sacral plexus supplies skin and muscles of the pelvic region. One of these nerves is the sciatic, which is the longest nerve of the body. It alone supplies the skin and muscles leading from the thigh to the foot, as anyone with sciatica knows only too well. Finally, the coccygeal plexus innervates the skin around the coccyx and anus.

Figure 12.2 Spinal nerves

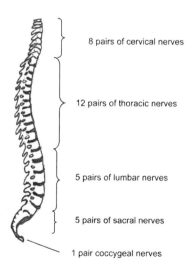

We have no voluntary control over the other part of the peripheral nervous system, which is the autonomic nervous system. This system is also connected to the central nervous system, and controls our internal organs. It has two major divisions, the sympathetic and parasympathetic nervous systems. A third division is called the enteric nervous system.

The sympathetic and parasympathetic nervous systems work together using opposite strategies. For example, if the sympathetic nervous system increases heart rate, the parasympathetic will decrease it. The sympathetic nervous system usually works under stressful situations, whereas the parasympathetic nervous system reacts under relaxed conditions. Activity of the autonomic nervous system depends on feedback received about the condition of the internal organs. The solar or celiac plexus is situated in the abdomen, and partly controls the spleen, pancreas, liver and stomach.

The sympathetic nervous system has networks of nerve cells situated in the thoracic and lumbar regions of the spinal cord. It affects the cardiovascular system by increasing heart rate and blood pressure. This diverts blood flow away from the skin and digestive organs and towards the heart and skeletal muscles. It affects the respiratory system by causing the muscle walls of the airways to relax, allowing greater amounts of air to enter the lungs when breathing in and increasing the breathing rate. The system delays the operation of the digestion by contracting its smooth muscle and inhibiting the production of digestive juices. The rate by which waste products are eliminated is slowed down through increasing muscle tone. When more energy is needed quickly, the system causes the liver to increase its conversion of glycogen to produce glucose. The system also inhibits the production of saliva, which is why the mouth and throat feel dry when we are stressed.

The parasympathetic system has similar plexuses of nerves which create opposite reactions when conditions are right. For example it will increase digestive absorption when the heart is in normal mode, and slows down breathing rate. It also causes the pupils of the eye to contract, and relaxes eyelids to produce the sleepy look when we feel tired.

The enteric nervous system controls the gastrointestinal tract throughout its entire length from the oesophagus (gullet) to the anus. Although it normally receives information from the central nervous system via the sympathetic and parasympathetic nerve networks, it can function autonomously. This is because of the extraordinary number of neurons it contains. It is sometimes known as a second brain, because of its neuronal complexity. The system's main function is to regulate and maintain optimal levels of fluids and nutrients.

How do nerve cells work?

Nerve cells are also called neurons. They are electrically excitable and transmit information throughout the nervous system. They vary in size, shape and function, and unlike other cells of the body, are generally unable to renew themselves. Therefore the resulting tissue damage, such as paralysis following an accident, usually remains permanent. Sensory neurons carry information relating to the outside environment to the brain and spinal cord along specific routes. They respond to light, smell, sound and touch. Motor neurons convey information from the central nervous system to the rest of the body. They can be activated either from a local site or by following instructions received from the brain via motor pathways, and cause muscles to contract or relax. They also affect glandular secretions.

Figure 12.3 shows the structure of a nerve cell. Each cell consists of a cell body containing a nucleus where proteins are synthesised. From the cell body are short extensions called dendrites which communicate to other target neurons. There are usually many dendrites to one neuron. A long tail called an axon extends at the other end of the cell at a small hillock, through which information flows away from the cell body. There is usually only one axon per neuron. Larger neurons are covered by a myelin sheath which insulates the axon. Myelin is produced by Schwann cells (after its discoverer). The sheaths are like strings of sausages, with the depression between each called a node of Ranvier (also after its discoverer). The signal jumps from one node of Ranvier to the next and speeds up transmission.

At the end of the axon are very small fibres called end feet or axon terminals. When the electrical impulse called an action

Figure 12.3 Structure of a nerve cell (neuron)

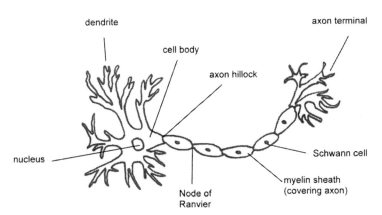

potential reaches this part of the cell, a chemical neurotransmitter is released which enables the signal to progress onto the next neuron along the pathway. Neurons receive information from many thousands of terminals. Figure 12.4 shows how the synapse sends information.

Between each neuron is a junction called a synapse and the tiny gap between the two is called the synaptic cleft. The transmitting signal reaches the pre-synaptic axon terminal which is triggered to secrete a chemical neurotransmitter. This is released into the synaptic cleft and attaches itself to special receptors in the post-synaptic membrane of a dendrite or cell body. The shape of each neurotransmitter exactly matches its receptor similar to the action of a key fitting a lock. Some well known neurotransmitters are serotonin, dopamine and adrenaline.

The trigger for the release of a neurotransmitter is a complicated set of electrical processes. When the action potential (signal) arrives at the pre-synaptic terminal, it

Figure 12.4 A synapse

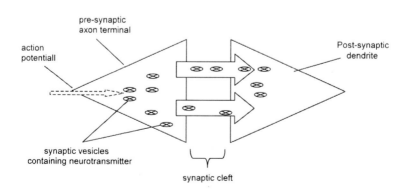

produces an influx of ions through voltage-gated channels. This alters the balance of positive and negative ions either side of the membrane. An ion is an electrically charged molecule dissolved in fluid within a cell. The resulting activity to restore the balance either excites to move the signal over to the next neuron, or inhibits any further action, if the signal is too weak. The larger the diameter of a neuron, the faster it travels. Impulses can be conducted to skeletal muscle at the rate of 130 metres per second. The slowest signals travel at about 0.5 metres per second.[1]

Glial cells supply a network of connective tissue between neurons, which provides structural support. These cells are only found in the nervous system. Unlike nerve cells they continue to renew themselves throughout life.

There are various other cells in the nervous system that perform vital functions. Astrocytes (so-called because they are shaped like a star) are a sub-type of glial cell, and link with axon terminals on the blood vessels of the brain and help to regulate the activities of neurons. They provide neurons with nutrients, such as glucose.

Whereas Schwann cells supply myelin around cells in the peripheral nervous system, oligodendrocyte cells myelinate axons in the central nervous system.

What causes seizures?

Seizures or convulsions can be produced by a trauma to the head following an accident, infection or tumour. This produces an intense abnormal electrical activity in the brain. Alcohol or drug abuse can also trigger seizures. Groups of neurons will unexpectedly and simultaneously discharge electrical impulses, causing an imbalance between excitatory and inhibitory neurotransmitters. The alteration in brain function is temporary. It may be that some congenital forms of epilepsy are due to mutations in ion channels.[2] There are various types of seizure, such as Grand Mal, Petit Mal or partial which produce symptoms ranging from loss of consciousness and involuntary body movements, to being unable to remember what happened during an occurrence. The symptoms depend which area of the brain is involved. In serious instances, the entire brain surface can be affected.

What causes the sensation of 'pins-and-needles'?

Paraesthesia is the medical name for the tingling sensation felt in the lower limbs. It is usually temporary. There are various causes, but the most common one is due to pressure on nerves. The peripheral nervous system sends information from the sensory nerves back to the brain and spinal cord. When it is pressed by being in a cramped or awkward position, the nerve is unable to process the information correctly, and the affected area becomes numb. When the pressure is relieved, a tingling sensation is produced, as pain signals are transmitted to the brain.

The 'pins-and-needles' sensation can also be created when a nerve becomes trapped or inflamed caused by an injury or infection.

A build up of placque on the walls of the arteries causes poor circulation, and can also produce the same symptoms. Under these circumstances, the feelings persist on a regular basis. The nerve cells have insufficient supplies of nutrients to function properly.

Disorders affecting the nervous system

As mentioned earlier, the brain has various protective devices, including the meninges and the blood-brain barrier. These usually prevent damage and injury. However, there are still some bacterial and viral infections which can invade its tissue and affect the nervous system generally:

Bacterial infections occur following an injury to the skull, by migration from other parts of the body to the brain via the bloodstream. Septicaemia and bacterial meningitis are infections caused this way.

Multiple sclerosis (MS) demyelinates nerve cells in the brain and spinal cord, leaving the nerve axons exposed and causing weakness of skeletal muscles.

Parkinson's disease is a degenerative disorder of the central nervous system. It affects movement and speech. The muscles become rigid and movements are slow. Arms and legs are also prone to tremors. The basal ganglia in the brain, fails to produce enough of the neurotransmitter dopamine, which causes a decreased stimulation of the motor cortex.

Spina bifida is a condition which presents defects in the development of the neural tube and the spinal cord during pregnancy. Although the cause is unknown, there appears to be a link with the amount of folic acid in the mother's system at the time of conception. Genetic or immune factors may also be implicated.

Viral infections can travel to the nervous system from other areas. Poliomyelitis enters the body through the digestive system via infected food. The virus travels through the bloodstream and invades the spinal cord. The shingles virus enters the respiratory system and can remain dormant in the spinal nerves of adults. These can become active many years later. Other slow forming viral infections are the human immunodeficiency virus (HIV) which affects the brain, and Creutzfeldt-Jacob disease (sometimes called Mad Cow Disease, because of its affect on herds of cattle). The condition leads to a progressive form of dementia and is fatal.

REFERENCES

[1] Waugh, A., and Grant., A.,(2006), *Ross and Wilson Anatomy and Physiology in Health and Illness*, Tenth Edition, Elsevier, p. 145

[2] McNamara, J.O., (1999), Emerging insights into the genesis of epilepsy. *Nature,* **399** (6738 Suppl.)., A15-A22

CHAPTER 13

PAIN and TOUCH

What is pain?
Why is pain good for us?
How does the sensation of pain reach the brain?
Why don't we feel pain in emergency situations?
What is referred pain?
What is phantom limb pain?
How do placebos work?
How do pain relief techniques work?
How do we experience different sensations of touch?
Which areas of the brain are involved with touch?

What is pain?

The nerve system relating to pain and touch is called the somatosensory system.

Pain can be described in different ways that can be understood by other people. We can express it by its strength, such as moderate, mild or agonising. The other way to explain it is by type. We say that the pain is throbbing, shooting, burning or sickening. It is always an unpleasant sensation. Pain can also describe an emotional feeling such as the ache of grief or loss.

Pain can also be categorised in terms of acute or chronic. Acute pain is experienced after an injury, infection or damage to the body. Chronic pain is long lasting and sometimes has no known cause. Backache, headache and arthritis are all types of chronic pain. Sometimes chronic pain can develop long after the initial cause has been successfully treated.

The first sign of pain is caused by tissue damage to areas close to special nerve fibre receptors that detect pain and are called nociceptors. These release chemicals that activate hormones such as prostaglandin, histamine and serotonin and create the conditions that elicit pain from the area. The skin has free nerve endings that respond to changes in temperature and chemicals.

Nociceptors are most numerous in the tissue layers of the skin. They produce short sharp pain in the damaged area caused by cuts and minor burns to the skin. There are less pain receptors in the joint regions and damage, such as a sprain in the ankle or wrist or a fractured bone, causes a dull ache in the surrounding area. Nociceptors are even more sparsely organised in and around the internal organs which is why we may experience less or no pain when they are diseased or damaged.

Why is pain good for us?

Pain is vital for survival. There are some people who are insensitive to pain, and they have an extraordinary amount of injuries. If we step on a sharp object and feel no pain, it will embed in the foot and become infected which can subsequently lead to serious disability affecting the whole leg. A patient in bed must be able to feel the discomfort of pressure points and turn at regular intervals to avoid bedsores.

Pain is a warning to remove, if possible, the affected area from danger. If we inadvertently touch a hot object, we instantly shift the hand away. Being in pain encourages us to rest, protect the injured area or obtain some medication for relief.

When we see someone in pain, we empathise and try to help, so the cry of pain is also a call for assistance. It is also a warning for us to adapt our behaviour and not get into the same situation again. But it doesn't always work. Drinking too much alcohol will result in a hangover headache the next morning. But we now have medications that will block the nociceptors and stop the pain. Therefore, we ignore the warnings and are likely to repeat our behaviour another time. Carrying the example to its extreme conclusion, there are few pain receptors in the liver, so we won't realise the potential damage until it's too late!

How does the sensation of pain reach the brain?

The brain itself does not have any nociceptors. That is probably because any damage is very serious and there is little that can be done to help. The pain of a headache is caused through nociceptors in the areas surrounding the brain and constriction of blood vessels.

When tissue is damaged, the chemicals that are secreted interact with nociceptors to produce an electrical signal (action potential). Pain is transmitted from peripheral areas to different regions of the brain through several ascending pathways of the central nervous system. The neuron axons synapse with other neurons and travel into the dorsal horn of the spinal cord. They then cross the midline of the spinal cord and travel to the medulla, midbrain and thalamus. The signals finish at the somato-sensory cortex in the brain.

The signals can travel fast through large diameter fibres called Aδ (A delta) fibres. These have myelinated sheaths and the action potential signal jumps between each sausage-like sheath. The speed of transmission to the brain means we can take fast withdrawal action when necessary, such as removing a finger from a hot object. The pain felt is sudden and sharp. A different, thin type of fibre is called a C fibre and these conduct the pain slowly to produce a dull ache. Figure 13.1 illustrates these receptors.

There are other descending pathways that transmit information from the hypothalamus in the brain back to the affected area via the spinal cord. They have the ability to reduce the level of pain, by inhibiting the release of chemicals. Figure 13.2 shows the main routes.

Figure.13. 1 Pain Receptors

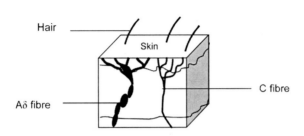

Why don't we feel pain in emergency situations?

People have been known to escape from serious incidents whilst suffering appalling injuries and loss of blood. In May 2003, a climber in Colorado amputated his own arm which was trapped under a large boulder, and hiked for five miles before being found.[1] Athletes have been known to finish the contest or game with broken bones. People in these situations have said they feel no pain until after they have reached safety, or completed the task. So, how do they overcome the pain of the injury?

The body produces its own analgesics, through the nervous system. This has evolved as a survival technique, when we need to overcome the pain barrier to deal with the situation. The perception of pain can be adapted if the brain considers a more important undertaking needs to be accomplished. Two scientists have proposed a gate control theory of pain, which seems to solve the question.[2]

The 'gate' is a junction that differentiates between the large Aδ fibres and the small C fibres. Large fibre inputs tend to close

Figure 13.2 Ascending and Descending Pain Pathways

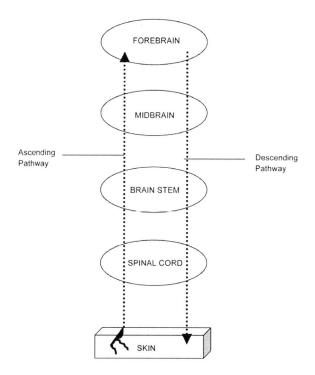

the gate, whereas small fibre inputs open it. The gate can also be closed by signals from the descending neural pathway, caused by information received from the brain. This inhibits the passage of small fibre pain signals to the brain. Figure 13.3 illustrates how this process works.

Opioids have the capacity to weaken pain levels. They bind to opioid receptors. This interferes with the transmission of pain signals, effectively closing the gate. Enkephalin is an opioid which is released at the endings of nociceptive neurons. Other neurochemicals, such as the hormone cholecystokinin, secreted at various spinal and brain regions will open the gate thereby increasing the experience of pain. It has a role in the high level of pain experienced during withdrawal from opioid drugs such as heroin.

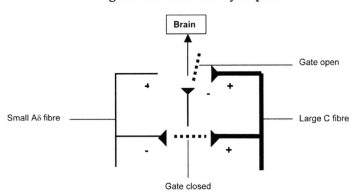

Figure 13.3 Gate Theory of pain

What is referred pain?

When a pain is experienced in a part of the body that is different from its origin, this is known as referred pain. For example, the pain of a heart attack can be felt in the neck, shoulder and back, but not the chest area. There have been several theories to explain the phenomenon, but most do not fully stand up to experimental scrutiny. Referred pain seems to be felt slightly later than the immediate pain sensed from local stimuli. This has led to the speculation that pain receptors in the inner organs also excite neurons in the dorsal horn of the spinal cord. The combined signals are interpreted by the brain as originating from regions of the skin. Another theory suggests that new receptive fields are created when the ascending pathways converge in the dorsal horn, and these produce referred pain. Another theory proposes that referred pain is caused by an accumulation of all the neural signals received through several converging pathways from different regions in the brain.[3]

What is phantom limb pain?

Pain is often felt in a missing limb, long after amputation and healing of the injured region. The sensations of pain indicate that the limb is still attached and that the brain has not dissociated from neural pathways that were previously in use when the limb was still part of the whole body.

Sensory nerve endings from the stump of the limb are misinterpreted by the brain. It is not just pain that is felt. Other sensations such as temperature, wetness and itching are also felt along the length of the missing limb. It has been suggested that the nervous system operates as a distinct entity on a personal level. People born with missing limbs can also experience phantom limb pain, which indicates that we are genetically pre-determined to expect sensations from all limbs.[4]

How do placebos work?

Placebos are substances that contain no active ingredient, and theoretically should be unable to alleviate pain or help cure any condition. A considerable amount of research and experimental evidence shows that in fact they work to a significant degree – the so called mind-over-matter phenomenon.

We learn from past experiences, or sometimes from the experience of respected others, what to expect under certain conditions. If we are prescribed a medication by a specialist practitioner working in a modern high-quality clinical environment, we expect it to help our symptoms. That expectation is sufficient in many cases to decrease pain.

The brain produces its own analgesic substances that will be secreted to the site of pain after swallowing a sugar-pill or receiving a saline injection. This is because the brain has been conditioned to expect relief. Opioids are substances that alleviate pain. Imaging scans have shown that the same areas of the brain are activated by opioids and placebos.[5]

Various brain imaging studies have taken place to examine the placebo effect on patients suffering with depression. One study showed changes in cerebral blood flow which were very similar to the changes after taking anti-depressant drugs.[6] Another study argued that the placebo effect was much more effective than actual treatment.[7]

The one condition that is necessary for a placebo to work is the *expectation* that it will. However placebos are not always successful unfortunately, otherwise our drug bills and side-effects would be substantially reduced.

The opposite effect can also be experienced. A nocebo response is the subjective prediction that the level of pain will be increased, simply because it is expected. Again, there is no active component to support the premise, as expectation is sufficient motivation to produce neural signals. Under certain circumstances, this response can be fatal, for example in the expectation of death after being placed under a voodoo spell.[8]

How do pain relief techniques work?

We often rub the site of pain, which seems to bring relief. This is because rubbing stimulates the large-diameter nerves which, if the gate theory is correct, will close the gate and prevent the signal from moving on towards the brain.

The body has its own analgesic system in the central nervous system and also in the peripheral areas in the skin. The central nervous system regions are in the spinal cord and inhibit the transmission of nociceptive neurons. Opioid receptors are activated when they bind to endorphins, which effectively close the gate and prevent further transmission of the pain signal.

Analgesic chemicals block secretions at the site of an injury to reduce inflammation. Some drugs block pain signals in the spinal cord.

Electrical stimulation through the skin triggers an analgesic reaction in areas of the brain by inhibiting nociceptive neuron activity between the pathways in the spinal cord to the thalamus. This form of pain relief has been used for many centuries. In AD46, the Roman physician Scribonius Largus records placing electric eels to painful areas to ease symptoms. He was physician to the Emperor Claudius. This type of treatment has been updated to a method of transcutaneous electrical nerve stimulation or TENS, which places electrodes around the affected area and produces electrical impulses.

Acupuncture by inserting needles to certain points in the body can also alleviate pain. The rotation of the needles stimulates the release of opiates. This may have a similar effect to TENS.

How do we experience different sensations of touch?

The skin contains four kinds of neuron receptors that are sensitive to different types of touch: Pacinian corpuscles, Meissner's corpuscles, Merkel cells or disks and Ruffini endings. Figure 13.4 illustrates the receptors. They are stimulated by sensations of pressure and vibration when the skin is pressed, and are distributed unevenly throughout the body. The most sensitive areas are the fingertips, lips and tongue, and the least sensitive is the back.

Figure 13.4 Touch receptors in skin and their receptive fields.

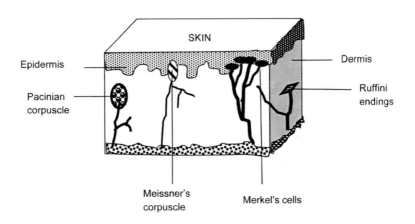

Pacinian corpuscles are quite long and thick. They are found deep in the dermis layer of skin. They are oval shaped and layered like an onion. They have a large receptive field and respond quickly to skin vibration within a range of 150–300 Hz. The layers act as filters detecting fast changing degrees of pressure and preventing further diffusion of pressure into lowest layer of skin.

Meissner's corpuscles are found in hairless, smooth skin, in fingerprint ridges, the lips and body orifices. They also respond to vibration, but at lower ranges. They have small receptive fields which adapt quickly to environmental changes. They inform the brain about the shape and feel of an object or the gentle touch of a kiss. They do not detect sustained pressure, which is why we do not notice the pressure of clothing on the body or the watch strap at the wrist.

Merkel cells or disks are oval shaped and are found in small clusters in the lower epidermis. They have small receptive fields and adapt slowly to detect steady pressure. Like Meissner's corpuscles, they are also found in areas where special sensitivity is required, such as fingertips, lips and tongue. They respond well to small points and edges. Merkel cells discriminate between the groups of dots on the Braille alphabet.[9]

Ruffini endings are spindle shaped with many branched nerve endings inside each capsule. They are found in the deeper dermis layer of skin over all the body, and are also slow adapters. They are sensitive to extended pressure on the skin, such as stretching when we move a limb.

Which areas of the brain are involved with touch?

The skin surface can be divided into sections called dermatomes. These carry nerves to specific spinal regions. Figure 13.5 shows the parts of the body that relate to each section.

Nerve fibres transmit signals from the skin to the somatosensory cortex of the brain via the relevant section of the dorsal root of the spinal cord, the medulla and the thalamus.

Those areas of the body that contain most nerve receptors have a relatively larger area of cortex with which to analyse the incoming signals. For example, the lips, tongue and hands have a wider band than arms or legs. This differentiation is usually described by a figure called a homunculus shown to a scale that is proportionate to its sensory space. Figure 13.6

Figure 13.5 Dermatomes

CERVICAL – Innervated by skin dermatome on upper limbs, neck, back of head

THORACIC – Innervated by skin dermatome on upper body (thorax)

LUMBAR – Innervated by skin dermatome on lower body (front)

SACRAL – Innervated by skin dermatome on lower body (back)

illustrates the concept. As you can see the thumb, fingers, lips and tongue cover a very large proportion of the somatosensory cortex compared to the rest of the body.

Figure 13.6 Somatosensory cortex and homunculus

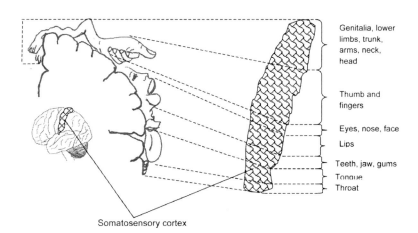

The somatosensory cortex is divided into six layers, as is the rest of the cortex. The cells in each column only respond to stimulation from a specific type of touch or pressure. Each column also only reacts to signals from a precise body location. However, the cortex is able to adapt to changing circumstances, and the areas can change as necessary. For example, a pianist or typist will have a larger than normal area that responds to nerve signals received from the fingers.

REFERENCES

[1] CNN.com/2003/US/Central/05/09/climber.amputation

[2] Melzack, R., and Wall, P.D., (1965), Pain mechanisms: a new theory, *Science*, **150**, pp. 971-979

[3] Arendt-Nielsen, L., Svensson, P., (2001), Referred muscle pain: Basic and clinical findings. *The Clinical Journal of Pain*, **17**, pp. 11-19

[4] Melzack, R., (1989), Phantom Limbs, the self and the brain (The D.O. Hebb Memorial Lecture). *Canadian Psychology*, **30**, pp. 1-16

[5] Petrovic, P., Kalso, E., Petersson, K.M., and Ingvar, M. (2002), Placebo and opioid analgesia imaging – A shared neuronal network. Science, **295**, pp. 1737-1740

[6] Leuchter, A.F., Cook, I.A., et al. (2002), Changes in brain function of depressed subjects during treatment with placebo. *The American Journal of Psychiatry*, **259** pp. 122-129

[7] Khan, A., Warner, H.A., and Brown, W.a., (2000). Symptom reduction and suicide risk in patients treated with placebo in antidepressant clinical trials: an analysis of the Food and Drug Administration database. *Archives of General Psychiatry*, 57, pp. 311-317

[8] Cohen, S.I., (1985). Psychosomatic Death: Voodoo Death in a Modern Perspective, *Integrative Psychiatry*, **Vol. 3, 1.**, pp. 46-51

[9] Rosenzweig, M.R., Breedlove, S.M., Watson, N.V., (2005), *Biological Psychology*, Fourth Edn.Sinauer Associates, Inc., p.230

CHAPTER 14

RESPIRATORY SYSTEM

Why must we breathe?
How does the respiratory system work?
Why are the lungs different sizes?
Why is there so much bleeding when the nose is damaged?
What is the function of nasal sinuses?
Is it true we can 'hear' through the nose?
What happens when we cough and sneeze?
How do we extract smells from the air?
Why are adenoids and tonsils more important for children?
What makes a new-born baby start breathing?
Is the brain involved in the respiratory system?
Why do we hyperventilate when stressed?
Why do we yawn?
Disorders affecting the respiratory tract

Why must we breathe?

The main function of the respiratory system is to exchange gases. All the cells of the body need a constant supply of oxygen for energy, and this is extracted from the air that is breathed in. As a result of chemical reactions, carbon dioxide is formed as a waste product and is breathed out. This trade of gases occurs in the lungs. On its way to the lungs through the respiratory tract, air from the environment is warmed or cooled to body temperature, and cleansed of unwanted materials such as dust or minute insects.

How does the respiratory system work?

Figure 14.1 shows the main respiratory organs. Air mainly enters the system through the nose, which has an internal

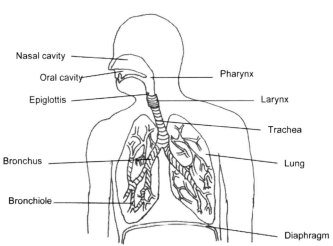

Figure 14.1 Respiratory system

division between two nostrils called the septum, with a passageway leading to the nasal cavity. This is lined with tiny hairs that are coated with sticky mucus which traps unwanted particles from the air. There are also ducts in the nasal cavity which form a canal to each eye, and drain tears. That is why the nose 'runs' when we cry. The roof of the nose also contains nerve endings which detect smells.

The nasal cavity leads into the pharynx, which consists of three separate parts. The nasopharynx has two openings (nostrils), each of which leads to the middle ear via the Eustachian tube. Unfortunately, throat infections can be spread to the ear via this tube. The oropharynx leads into the oral cavity, and finally the laryngopharynx is the food passage and leads into the oesophagus or gullet. Food passes through the pharynx on its way to the oesophagus and air passes through it towards the trachea, also known as the windpipe. To prevent food entering the trachea, a piece of cartilage called the epiglottis acts as a lid and covers the larynx during swallowing. If small particles accidentally enter the larynx, they irritate the lining which causes coughing or choking to eliminate them through the mouth.

Leading off from the pharynx is the larynx or voice box (see chapter on Language). Sounds are made by expelling air from the lungs. The larynx moves up to cover the opening to the pharynx whilst swallowing.

The passageway of air continues from the larynx into the trachea. It is made up of 'C' shaped rings of cartilage and elastic tissue which allows the tube to remain stable when the head moves. Muscle and connective tissue at the back of the trachea complete the circles. The trachea extends about 12 centimetres towards the diaphragm and divides into two bronchi (singular, bronchus), each of which extends into the left or right lung. Tiny hairs in the mucus membrane lining of the trachea fan particles up towards the larynx, where they can be coughed out through the mouth, or swallowed into the digestive system.

The two bronchi are smaller versions of the trachea. The left bronchus is narrower and shorter than the right and divides into two branches, each of which enters a lobe of the left lung. The right bronchus divides into three branches, which enter lobes of the right lung. These branches continue to subdivide into ever smaller tubes called bronchioles, and the walls change their structure from cartilage into smooth muscle. The walls are thicker and respond to nerve signals. They regulate the amount of air entering the lungs, by an involuntary contraction or relaxation of the muscle walls.

The lungs are soft and spongy and have a protective dual-layer covering called pleura. Between each layer is the pleural cavity containing a fluid which prevents friction when breathing. The pleura can become infected and inflamed, causing the condition known as pleurisy. The lungs are composed of tissue, consisting of alveoli or air sac cells. There are about 300 million alveoli in the lungs. They are tiny grape-like bunches of cells which are the final result of the sub-division of bronchioles. Alveoli are surrounded by a network of tiny blood vessels called capillaries. The exchange of oxygen and carbon dioxide occurs between the semi-permeable membrane separating the alveoli and capillaries. Figure14. 2 shows the structure of alveoli

Figure 14.2 Alveoli

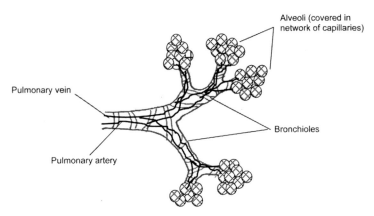

There are two kinds of respiration: internal and external. External respiration is the exchange of gases between air in the alveoli (singular alveolus), which originated in the atmosphere, and blood in the capillaries. When the blood arrives at the lungs from the heart and surrounds the alveoli, it contains high levels of carbon dioxide and low levels of oxygen. As air arrives at the alveoli with oxygen, this creates a pressure difference with the amount of oxygen in the blood capillaries. There is a consequent diffusion of gases across the membrane to equalise the pressure and concentration gradients of each gas, as shown in Figure 14.3.

Internal respiration is the exchange of gases between the blood in the capillaries, and the body cells. There is a high level of oxygen in the capillary blood and this diffuses into body tissue cells. At the same time, carbon dioxide diffuses from the cells into the bloodstream. Figure 14.4 shows a diagram of the process.

Figure 14.3 External respiration

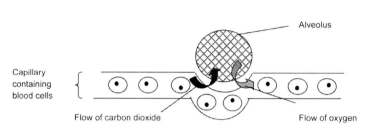

The main muscles involved in breathing are the intercostal muscles between the ribs and the diaphragm. These muscles contract when we breathe in, and pull the ribs towards the first fixed rib. The central tendon of the diaphragm muscle contracts, and pushes down the abdominal cavity. Both of these actions expand the chest area when we breathe in. This causes a decrease in pressure in the lungs, and an increase in pressure from incoming air that flows into the lungs, causing them to expand. Hiccups are triggered by an involuntary repeating spasm of the diaphragm, which creates a rush of air into the lungs. The epiglottis closes resulting in the 'hic' noise.

When we breathe out, the intercostal muscles and the diaphragm relax to their normal positions. The lungs have elastic recoil, and with increased pressure most of the air is pushed out. There is always some air left behind which stops the lungs collapsing. Interestingly, laughter is a series of short repeated breaths out.

In an average resting adult there are about 15 respiratory cycles per minute.[1]

Figure 14.4 Internal respiration

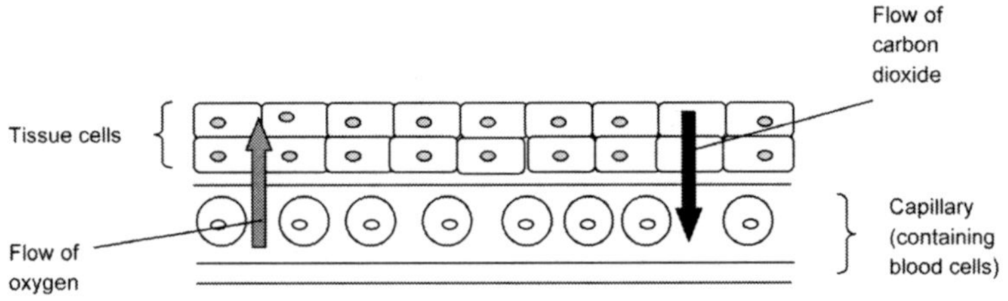
Flow of
carbon
dioxide
Tissue cells
Capillary
(containing
blood cells)
Flow of
oxygen

Why are the lungs different sizes?

The right lung is divided into three lobes. The left lung is smaller and is divided into two lobes. This is because the heart occupies space slightly to the left of the middle of the thorax. Figure 14.5 shows a diagram of the lungs and heart, and as can be seen, the heart squeezes the left lung out of position.

Figure 14.5 Position of heart relative to lungs

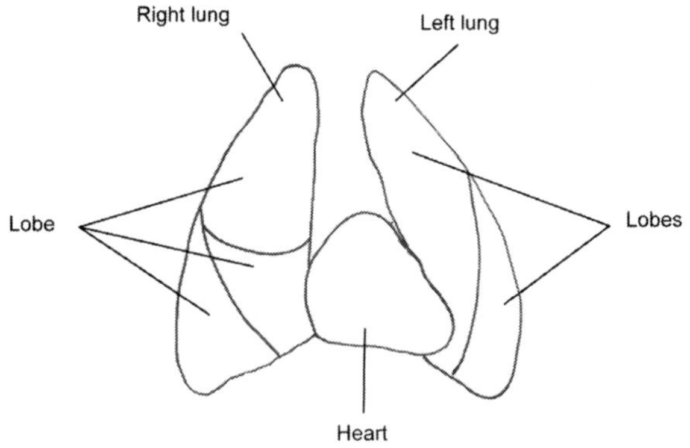
Right lung
Left lung
Lobe
Lobes
Heart

Why is there so much bleeding when the nose is damaged?

Nosebleeds are caused by tearing of the mucus lining of the nose. This lining has a huge amount of blood vessels, and when they are damaged through infection, injury or high blood pressure, the walls split. Because there are so many minute blood vessels, there is consequently a greater amount of blood loss.

What is the function of nasal sinuses?

The sinuses are four pairs of cavities in the bones of the face and skull. Each part of the pair lies on opposite sides of the face. They contain air and have tiny openings into the nasal cavity. Their function is to decrease the weight of the facial bones and increase the resonance of the voice. Infection or allergies can cause sinusitis, and causes inadequate drainage through the cavities, triggering painful inflammation and blockage.

Is it true we can 'hear' through the nose?

From each side of the nasopharynx, where it joins the pharynx, there is an auditory tube called the Eustachian tube. This leads directly to each middle ear. It is made up of bone and cartilage and air travels at atmospheric pressure to each ear drum. The tube is normally closed, but can open to allow air through to equalise pressure between the middle ear and the atmosphere. This process is familiar to people who travel on high mountain roads or in aircraft, when they experience a small popping sound in the ears. Also when this happens, the sound of your own voice can seem to echo and become louder. Hearing clearly relies on air at atmospheric pressure on each side of the ear drum.

What happens when we cough and sneeze?

A cough is a spasmodic contraction of the abdominal and respiratory muscles, which results in the violent release of air from the lungs. As the air rushes through the larynx, it produces sounds, and expels mucus or other unwanted substances. Sensory nerves in the upper respiratory tract are sensitive to irritation, and nerve impulses send signals to the brain which starts the process.

A sneeze is a convulsive explosion of air from the nose and mouth, which forces unwanted particles out of the nose. Expelled air can reach speeds of about 150 kilometres per hour. It is caused by irritation to the mucus in the nose, and is a reflex action.

Why are adenoids and tonsils more important for children?

Adenoids and tonsils are masses of lymph glands, and form part of the immune system. They trap viruses and bacteria, and produce antibodies. The adenoids are an additional part of the immune system in children, and are more important up to the age of seven. As the immune system matures, they are no longer needed as a back-up and shrink away. The tonsils remain throughout life, and can be seen as two fleshy lumps at the back of the mouth. They can become infected and enlarged and result in tonsillitis. The removal of adenoids and tonsils that are causing regular problems appear to have no adverse effect on the remainder of the immune system. Like the appendix, they may be vestigial structures which have lost their original function through evolution.

What makes a new-born baby start breathing?

As the foetus develops, the oxygen and nutrients needed are received through the blood circulatory system, via the placenta and the umbilical cord link with the mother. There is an opening between the main arteries at the heart which allows foetal blood to bypass the lungs. Meanwhile the lungs are filled with a fluid surfactant which is made in the cells of the respiratory tract and released into lung tissue. The effect is to lower surface tension in the airways, and keep the alveoli open. Ultrasound scans have shown that a foetus makes regular breathing movements to move the fluid in and out of the lungs.

When the baby is born, the opening at the heart is no longer needed and closes up, allowing the lungs to take over. They expand with air as the baby takes its first breath, and the fluid is pushed out of the body.

Is the brain involved in the respiratory system?

Respiration is mainly regulated by the autonomic nervous system in the brain stem at the base of the brain. Although breathing is involuntary, we do have some control to stop, slow down or speed up the respiratory rate. We need to have this control to be able to speak or sing or cry, which interrupts the normal breathing cycle. But this voluntary control is only temporary. When we hold our breath, we are starving the brain of oxygen and creating a build up of carbon dioxide. Chemical receptors in the brain detect levels of carbon dioxide in the blood. Too much increases the rate of breathing, and too little slows it down. The autonomic nervous system will take over to restore breathing, to prevent loss of consciousness.

Why do we hyperventilate when stressed?

When we experience very stressful situations, we tend to breathe very quickly, sometimes called over-breathing. This unbalances the system by increasing the amount of oxygen in the blood stream and too little carbon dioxide. This is because atmospheric air contains much more oxygen (about 21%) than carbon dioxide (about 0.4%). Under normal conditions, we breathe at appropriate rates to supply the body with the correct proportions of oxygen and carbon dioxide. The gases in the alveoli of the lungs are in approximately equal proportions. Breathing more quickly exchanges more of the alveolar gas with air, drawing in more carbon dioxide from the blood stream. Low levels of carbon dioxide in the blood causes the blood vessels in the brain to constrict, resulting in reduced blood flow to the brain, which creates a light-headed feeling. This can be corrected by breathing into a paper bag, as this increases the levels of carbon dioxide in the air in the bag as we breathe out, and then inhale more carbon dioxide.

Why do we yawn?

There is no definitive answer to this question, as various

studies seem to refute the usual theories, leaving a puzzling question to the phenomenon.

We yawn when we are tired, stressed or bored. The old hypothesis was that yawning was caused by an excess of carbon dioxide and lack of oxygen in the blood. This stimulates neurons in the brain stem to trigger the yawn reflex. A wide-open mouth inhales a larger-than-normal volume of air which brings oxygen to the lungs and bloodstream. The improved circulation makes us feel more alert. However studies have indicated this is not true.[2]

Another theory is that yawning is a form of stretching, which increases blood pressure and heart rate and flexes muscles and joints.

A curious effect of yawning is that it appears to be contagious. Most of us have at times yawned soon after seeing someone else yawn. Specialised nerve cells called mirror neurons located in the brain are activated when we are involved in imitation or learning. One idea is that the yawn is linked to this system.[3] An alternative suggestion is that the yawn is an evolutionary form of herd communication.[4] It signals to the group that it is time to coordinate a sleep routine.

Disorders affecting the respiratory tract

The upper part of the respiratory system leading to the division of the trachea into the two bronchi is subject to bacterial and viral infections, some of which have been mentioned under the sections relating to tonsils and sinuses. Colds and influenza are infectious diseases causing a discharge of mucus from the nose, sore throat and a raised temperature. The virus can also inflame the pharynx causing **pharyngitis**. **Laryngitis** can also develop from this same group of germs. **Diphtheria** is another bacteria affecting the pharynx and trachea causing a thick membrane to form and obstruct the tract. The disease can have a harmful affect on the heart and skeletal muscle, as well as the liver, kidneys and adrenal glands.

Hay fever is an allergic reaction to certain inhaled particles such as pollen or dust mites.

Bronchitis is an infection of the bronchi and can develop from a cold or influenza. The virus allows bacteria already in the system to breed.

Asthma is another condition which inflames the bronchial muscles. A thick mucus forms and the glands enlarge, reducing the flow of air in the tract. Air is only partially exhaled, leaving the lungs hyper-inflated and causing a wheeze. There are two forms of asthma: one affects the young, and the other affects adults. The childhood type is caused by inhaling antigens. These produce antibodies in the blood vessels of the bronchi. The antibodies generate histamine, which causes an excessive amount of mucus. The respiratory tract narrows, creating breathing difficulties. The type that affects adults, results in inflammation of the bronchi. Triggers can be dust mites, animal fur, pollen or tobacco smoke.

There are several conditions affecting the lungs:

Emphysema is a condition that is caused by exposure to toxic chemicals and long-term cigarette smoke. Lung tissue loses its elasticity and the walls of the alveoli collapse making it difficult to breathe.

Pneumonia is an infection of the alveoli which become inflamed and flood with fluid. It can also develop from a complication of another illness such as lung cancer or alcohol abuse.

Tuberculosis is caused by microbes which colonise the bronchioles in the lungs. They form small tubular structures which prevent the immune system from destroying them. Over time the infection can spread to other areas of the body, or can be halted if the immune system recovers.

The lungs can also be affected by conditions in the work-place:

Asbestosis is triggered by inhaling blue asbestos fibres, which destroys lung tissue.

Silicosis is caused by dust containing silicon raised and inhaled when quarrying metals or stones such as granite or slate. The toxic particles accumulate in, and inflame the alveoli.

Pneumoconiosis is produced from coal dust in the lung, and inflames the lung tissue.

Cystic fibrosis is an inherited disorder affecting babies. It is caused by an abnormal gene. Both parents must carry the gene for it to affect the baby. Overproduction of a thick mucus results in frequent lung infections. The gene also creates excessive sweat containing more salt than the body can cope with. The other effect is that the digestive system is unable to process food satisfactorily.

REFERENCES

[1] Waugh, A and Grant, A., (2006), Ross and Wilson Anatomy and Physiology in Health and Illness, Churchill Livingstone Elsevier Limited, p. 252

[2] Provine, R.P., (2005) "Yawning", pp 532-539, *American Scientist*, **Vol 93, No. 6** (http://www.americanscientist.org/template/Assetdetail/assetid/47361)

[3] Ramachandran, V.S., (2006) Mirror Neurons and imitation learning as the driving force behind "the great leap forward" in human evolution (http://www.edge.org/3rd_culture/ramachandran_pl.html)

[4] Schürmann et al.(2005) Yearning to yawn: the neural basis of contagious yawning. *NeuroImage* **24(4)**, pp. 1260-1264

CHAPTER 15

SEX AND REPRODUCTION

What are the mechanics?
Why do we need two sexes to reproduce?
How is the brain involved in sexual attraction?
What is the difference between heterosexual male and female attitudes to sex and reproduction?
Do humans have attracting pheromones?
How does Viagra® work?
Is there a gay gene?
How does alcohol or drugs affect sex and reproduction?
What is cloning?
Disorders affecting the reproductive system

What are the mechanics?

All cells in the human body, apart from sperm and ova, have 23 pairs of chromosomes. A chromosome is a long strand of deoxyribonucleic acid (DNA) containing proteins and other molecules. The DNA consists of all the genetic information for development. Humans have two types of chromosomes: autosomes and sex chromosomes. The autosomes replicate by a process of cell division called mitosis. During this process the cell makes an identical copy of each chromosome and the result is two new cells, each containing a complete set of 46 chromosomes (see figure 15.1). Each new cell contains the same genetic information as the 'parent' cells.

The sex chromosomes are X and Y. The human female has two X chromosomes and the male has one X or one Y. The nuclear division of the gamete cells involved in reproduction is caused by a different method of division called meiosis. When a sperm penetrates the egg at fertilisation, its nuclei combines

Figure 15.1 Mitosis

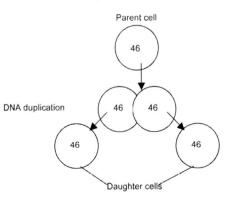

and forms the nucleus of a zygote. Both genders have the remaining 22 pairs of autosomes. All carry different genes. Meiosis occurs in the gonads, which are the ovaries in females and the testes in males (see also chapter on Atoms, Cells and Genes).

Unlike mitosis, meiosis halves the number of chromosomes resulting from the division of the parent cell. Each gamete contains only 23 chromosomes, one from each of the 22 pairs of autosomes, plus either an X or Y chromosome. When the eggs are fertilised, the combination results in a total of 23 pairs, one half coming from the male sperm and the other from the female egg (see Figure 15.2). Most of the cytoplasm of the cell is directed into that part which will nourish the egg. The polar bodies shown in the diagram are non-functional parts of the cell with tiny amounts of cytoplasm. They appear to only assist in the process of cell division, following which they disintegrate.

Figure 15.2 Meiosis

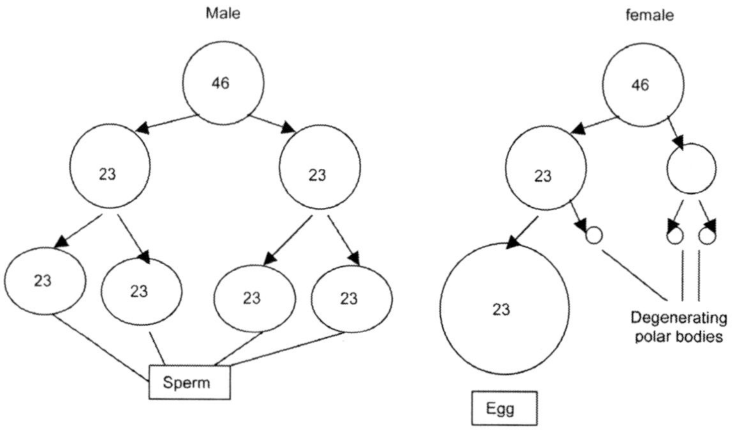

The result is that each fertilised egg has a unique combination of genes from both parents. Following fertilisation, the egg develops using the process of mitosis, retaining the normal numbers of chromosomes.

The female egg always has an X chromosome. If a penetrating sperm also has an X chromosome, the resulting foetus will be a girl. If the sperm has a Y chromosome, a boy will develop. A gene on the Y chromosome called the Sex Determining Region Y (SRY) produces an enzyme, which causes the sex organs (gonads) to become testes. In the absence of this gene, ovaries will develop instead. Up to the third month after conception, the sex organs of the foetus are bi-sexual. After that, one set continues to develop and the other decays.

As shown in Figure 15.3 the female sex organs include the fallopian tubes, the uterus, the vagina and the ovaries. The fallopian tubes are about 10 cm long and extend each side of the uterus. There are finger-like projections at each end that collect the follicle to begin its journey towards the uterus. The uterus is a muscular hollow organ. The neck of the uterus is called the cervix, which protrudes through the wall of the vagina. The vagina is a muscular, ridged tube, which is kept moist by secretions from the cervix. The ovaries are attached to the uterus and are each about 3 cm long, 2 cm wide and 1 cm thick.

Male sex organs include the penis, the epididymis, vas deferens, seminal vesicles and the prostate (see Figure 15.4). The scrotum is a pouch that contains one testicle. The testes are equivalent to the ovaries in females. They produce two kinds of hormone: a peptide, which defeminises to prevent the growth of female reproductive system, and steroid

Figure 15.3 Female reproductive organs

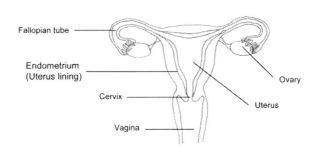

hormones called androgens which masculinise, one of which is testosterone. Androgens stimulate and control the development and maintenance of masculine characteristics.

Sperm is produced in the testes and stored in the epididymis. The head of the sperm contains the nucleus filled with DNA, as well as enzymes able to penetrate the outer layers of the egg. The seminal vesicles are glands which store fluid that will form the major part of semen ejaculated at male orgasm. The prostate gland secretes a fluid that also contributes to the formation of semen, and gives it an opaque appearance. This fluid contains an enzyme that thickens the semen to enable it to stay close to the cervix. The vas deferens is a duct connecting the epididymis to the seminal vesicles.

The male urethra is a pathway that has a dual function. It allows urine to pass out of the body from the bladder, and semen to be ejaculated. When stimulated, the penis becomes erect and engorged with blood, which is an essential process for sexual intercourse.

At puberty, around the age of twelve, secondary sex

characteristics begin to appear. In girls, breasts start to develop, hips widen and fat is redistributed to the hips and buttocks. In boys the voice deepens and a facial beard and sometimes chest hair start to grow. Also underarm and pubic hair begin to grow in both sexes.

Figure 15.4 Male reproductive organs

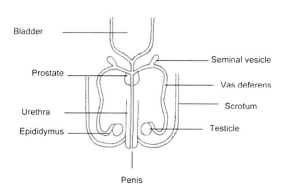

The ovaries produce oestradiol from a class of hormones called oestrogens. Oestradiol develops the breasts, and causes the lining of the uterus to grow. The testes produce the androgen testosterone, which promotes male features. These gonadal hormones halt further growth of the skeleton.

Humans have a reproductive (menstrual) cycle, lasting about 28 days. The cycle begins with the secretion of the hormone called follicle-stimulating hormone (FSH) that stimulates the growth of ovarian follicles around each egg. Usually one is produced at each cycle, sometimes two resulting in non-identical twins. In very rare cases, triplets and quads may result from additional follicle production. Identical twins develop from the division of only one fertilised egg. As they mature, the follicles secrete oestradiol, causing the lining of the uterus (endometrium) to thicken in readiness for implanting of the fertilised egg. Ovulation is caused by a surge of luteinizing hormone (LH), which releases the egg, and enters the fallopian tube. It then travels towards the uterus and meets sperm. The fertilised egg attaches itself to the wall of the uterus, where it will be nourished, and begin the process of multiple division. If the egg is not fertilised, the lining of the uterus will be released during menstruation.

Why do we need two sexes to reproduce?

Natural selection has evolved an advantage in reproducing from two different individuals. If humans were able to reproduce asexually, the children would have 100% of the genetic make-up from one parent. Any genetic dispositions towards inherited diseases would be passed down. Eventually, the line of inheritance would become extinct when harmful mutations would kill off individuals before they were able to reproduce.

Sexual reproduction ensures that every individual has a unique combination of the genes from two parents. These will often cancel out any inherited disorder from one parent only. Also advantageous characteristics will adapt to changes in the environment and, through sexual reproduction, will be gradually incorporated into the majority of the population. Such advantages throughout our evolutionary history include physical changes in the location of the larynx in the throat, which gives us the ability for speech. Another is the adaptation of an opposable thumb that allows us to grasp, use and make a variety of small and delicately shaped objects. This followed on from another adaptation, the ability to walk on two feet only, freeing the hands for other duties. These attributes are the direct result of sexual reproduction.

How is the brain involved in sexual attraction?

Sexual motivation is triggered by both external stimuli, such as being attracted by aspects of the physical appearance of another person, or looking at pornographic material. There are also internal factors such as hormones, which are transported in the bloodstream. When they reach certain areas of the brain, neuronal activity triggers sexual behaviour.

In adolescent boys, testosterone levels correlate with the frequency of sexual thoughts and actions. PET scans have shown that the amygdala in the brain becomes activated when a person is sexually aroused. The amydgala stores memories associated with emotional events.

Men who have been castrated have very low levels of sexual desire, because testosterone is no longer secreted in the testes.

The adrenal glands are situated just above the kidneys. Women, who have had them surgically removed for health reasons, also showed reduced levels of sexual desire. This can be restored by injections of testosterone.

The hypothalamus is connected to the pituitary gland at the base of the brain. It releases a gonadotropin-releasing hormone (GnRH), and causes the pituitary gland to release LH and FSH triggering sexual activity.

The Coolidge Effect demonstrates the role that novelty plays in sexual attraction. The story is that the late President of the United States and his wife were visiting a farm, and taken on separate tours. Mrs. Coolidge was told that a particular rooster copulated many times a day, and said "Tell that to the President". The President was told that the rooster used a different hen each time, and said "Tell that to Mrs. Coolidge".[1] Tests have shown (usually with rats – but can apply to humans too!) that even when sexually sated, renewed interest can be achieved through a new association.

Carrying out experiments on humans is considered unethical, so most evidence comes from studying people with brain problems. Genetic and cultural traditions may form neural networks in the brain which become models for sexual behaviour. Patients suffering damage to the prefrontal cortex seem to develop a much less inhibited attitude to sex.[2]

Damage to the spinal cord can prevent neuronal messages being transmitted between the brain and the penis, although an erection can still occur because the reflex pathways remain undamaged.[3]

The right hemisphere in humans seems to be the dominant side of the brain for triggering sexual desire.[4] This may be because the right side relates to emotional stimuli more than the left hemisphere.

What is the difference between heterosexual male and female attitudes to sex and reproduction?

Human adults are unique amongst mammals in having clearly

visible physical differences between the sexes, especially amongst adolescents in early maturity. In women, there is a redistribution of adipose fat to the breasts, thighs and buttocks. Facial and chest hair and a general muscularity in the arms and thorax, denote masculinity. As these differences are not necessarily related to a reproductive period when a woman is pregnant or breast-feeding, they must be signals of sexual attractiveness.

Physical attraction depends upon looks, smells, and sounds. The person should look healthy and well proportioned, should smell fresh, and have a pleasant voice. These qualities apply to both men and women.

The supply of male sperm is almost boundless and in that respect it is very cheap to produce. Therefore men can afford to spend it without restraint. Only cultural restrictions prevent uninhibited sexual activity. But the male is also looking for a mate who will be able to bear his children. Therefore she should be reasonably young and fertile and show physical signs that she can successfully make good use of his sperm. A small waist emphasises that the breasts, hips and buttocks have a good supply of fat, induced by high levels of oestrogen. Long hair and nails, and unblemished skin all indicate good health, and are desirable attributes. Women will buy cosmetics to enhance these features, such as false nails and hair extensions and facial make-up.

However, the supply of female eggs is limited. All available eggs, from adolescence to menopause are already in place in the ovaries at birth. Normally only one egg is released during each monthly cycle. It is a very expensive resource and a woman must be very careful about the choice of a partner, to be the father of her children. Human children take many years to mature, and women devote a great deal of their time caring for their offspring. Ideally the father should also be around to help out.

Therefore as well as looking for signs that he is healthy a woman looks for other signs in a man, if she wants to have children. She doesn't necessarily look for youth, but on his ability and willingness to stay and provide for them, both physically and financially whilst the children are young.

Since the advent of reliable methods to control conception, men and women have probably for the first time in history been able to make choices about sexual encounters that did not involve pregnancy and parenthood. In Western societies, attitudes towards unmarried mothers and couples have softened. The differences between men and women's approach towards sex have consequently blurred the evolutionary imperative. Nevertheless, the basic strategies remain in place, towards the ultimate goals of reproduction and genetic inheritance.

Do human have attracting pheromones?

Pheromones are scented sex hormones. Animals have a vomeronasal organ (VNO) in the nose that detects pheromones in the opposite sex and leads to mating. In humans this organ regresses during foetal development. There is a vestigial pit, which some researchers have concluded could still be functional at an unconscious level. However, there is no evidence to date of a neural connection between the VNO and the brain, so it is unlikely that pheromones can affect human sexual behaviour.[5]

However, they may be a minor contributor to human communication. Studies have shown that groups of women living in close proximity tend to synchronise their menstrual cycles.[6] It could be that in evolutionary terms if females in a group all became fertile simultaneously, this could maximise reproductive success and other community functions such as moving or hunting could be carried out at other times.

The power of human pheromones has often been exaggerated by advertisers, keen to exploit the cosmetic market with claims of sexual irresistibility.

Androstenol is a male scent produced by fresh sweat. It is appealing to women, especially when they are ovulating. However, when it is exposed to air, it oxydises, becomes stale, and turns into adrostenone, a powerful odour which is definitely unattractive to women.

Women wear perfumes to enhance their appeal. However, the confidence this brings, may alter behaviour, and be the reason why they are more attractive, and not the smell itself.

As pheromonal action is unconscious in humans, there may be many other social reasons for mutual attraction, which have nothing to do with smells. Our sense of smell has radically reduced throughout history, probably because we rely much more on visual signals to determine behaviour.

How does Viagra® work?

The main ingredient in Viagra® is sildenafil citrate. Originally produced as a treatment for high blood pressure, it was subsequently found to also help impotent men regain an erection. It enhances the action of nitric oxide in the penis.

Nitric oxide binds to receptors that increase levels of a nucleotide (chemical compound) called cyclic guanosine monophosphate (cGMP). These receptors cause the muscles of the penis to relax, allowing an increased flow of blood, which produces an erection.

There are common side effects such as sneezing, headaches, palpitations, and in some cases prolonged erections – which could prove counter-effective.

Viagra® has also been taken as an aphrodisiac, but there is no scientific evidence to back up the effect.

Is there a gay gene?

Humans are the only species with some members known to be exclusively homosexual. Testosterone levels are the same for homosexual and heterosexual men, although some studies indicate there may be higher than average levels in homosexual women.[7] Post-mortem examinations of homosexual men have shown that some areas of the brain are different in size, especially in a bundle of fibres that conveys information between the right and left hemispheres.[8] However, this may be due to the effects of the causes of their deaths. It may also be that homosexual activity restricts or enlarges these areas.

The differences in brain structures may be caused by unusual patterns of hormones presented to a foetus at critical stages of development. These could be caused by stress or trauma or heredity factors. But there is little actual evidence to substantiate these claims.

A famous study by Simon LeVay, indicated a genetic bias towards sexual orientation.[9] It's a bit of a puzzle, because a gay gene should not have survived throughout many generations, as homosexual adults would be less likely to reproduce. However, it could be that bisexual adults carry the gene and have passed it down to future generations.

Currently, evidence for a gay gene appears weak. Environmental factors may prove more reliable. A word of caution though –the previously held views that domineering mothers or violent fathers caused their children to become homosexual has been entirely discredited.

How does alcohol or drugs affect sex and reproduction?

In small doses, alcohol releases inhibitions, and leads to the assumption that it can enhance sexual performance, but this is untrue.

In men, excessive use of alcohol has a detrimental effect, by inhibiting the production of testosterone and damaging the neurons that cause an erection. Ethanol is an alcohol used in drinks and inhibits the actions of the hypothalamus and pituitary glands (see previous paragraphs and also chapter on Drugs and Addiction). The results are shrinkage of the testes and the penis, and a low sperm count. Alcohol also inhibits the metabolism of Vitamin A, which is essential for sperm development.

Excessive use of alcohol by women can lead to a failure to ovulate and general menstrual problems. It has a particularly damaging effect on the foetus. Babies can be born in severe pain showing acute withdrawal symptoms. There are also birth defects, including facial abnormalities, retardation, low birth weight, which carries its own risks, and defects of the central nervous system. Alcohol also affects sexual performance as it dehydrates the system and leads to loss of lubrication in the vaginal canal.

The side effects of drugs also have a detrimental effect on sex and reproduction. In the 1960's pregnant women taking the drug thalidomide to relieve the symptoms of morning sickness in the early stages of development, led to the malformation of limbs in many babies.

Nicotine, marijuana and opiates such as morphine, all reduce testosterone levels, and long-term use may lead to impotence. In men, other symptoms include blockage of the vas deferens and epididymis (see first paragraph) and causes smaller orgasm.

It is generally accepted that women can safely drink 14 units of alcohol per week and men 21 units. Alcohol is dispersed through the body in water and women have less water than men, as they have a lower body-weight. Therefore the alcohol stays in the body for longer periods. Half a unit is equivalent to a glass of beer and a glass of wine contains about one and a half units of alcohol.

All-in-all, it would appear that alcohol and drugs have a very negative effect on sex and reproduction.

What is cloning?

Cloning involves making duplicate copies of genes or cells, for reproductive or therapeutic purposes. The experiment, first carried out successfully in 1997, resulted in the birth of Dolly the sheep, using a technique called somatic cell nuclear transfer.[10] A somatic cell is any cell that is not sperm or egg. The nucleus of a somatic cell containing its DNA is removed and inserted into an egg cell which has had its nucleus removed. The cell is then stimulated with an electrical charge and starts to divide, eventually forming a blastocyst, which is an early-stage embryo containing about 100 cells. Its DNA will be almost identical to the first cell.

This procedure can be used in stem cell research. However, with current knowledge the process has severe limitations. There is a very high loss of eggs, and concerns about the health of the resulting progeny. Dolly the sheep was produced after using 277 eggs, from which only 29 embryos were initially viable and only Dolly survived.[11] Although theoretically the process could be used to clone humans, there are many ethical concerns and as mentioned above, it is highly unlikely this could be carried out successfully using this technique.

Disorders affecting the reproductive system

In males:
Gonococcal urethritis is a urinary infection caused by the gonorrhoea bacterium. It is spread through sexual contact. The symptoms are discharge from the penis, and a burning

sensation when urinating. There are other forms or urethritis which can spread from the bladder. The infection can be spread to other parts of the male reproductive system. It is treated with antibiotics.

Gynaecomastia is a condition where the breast tissue grows abnormally. It is caused by abnormal hormone levels, particularly of oestrogen, or by cirrhosis of the liver.

Hydrocele is a swelling in the scrotum caused by an accumulation of serous fluid. It is most common amongst the middle aged, and is usually benign. It may be caused by an injury, infection or in rare cases by a tumour.

Orchitis is an infection which causes acute inflammation and swelling of the testis. It can be spread from an infected urethra or bacteria. Bacterial infections can be treated with antibiotics. Another common cause is by the mumps virus, which in severe cases can lead to sterility.

Prostatitis is an inflammation of the prostate gland, caused by a bacterial infection, which may have spread from the urethra or bladder. There is increased frequency in and pain during urination. It is treated with antibiotics but can recur.

Testicular tumours usually occur in young men. The tumour causes a swelling in one testis and can spread if untreated. Treatments in the early stages are very successful.

Undescended testis (cryptorchidism) is a condition that usually only affects one testis. The testes are developed in embryos in the abdominal cavity and normally descend into the scrotum shortly before birth. A testis may also descend soon after birth. An undescended testis will be incapable of producing viable sperm.

In females:

Cervical carcinoma has various stages and is usually identified through a cervical smear test, and can be treated in the early stages. If undetected, it can spread to the uterus, vagina, bladder and rectum and on to the liver, lungs and bones via the bloodstream. It can take up to 20 years to develop. One type of carcinoma appears to be caused by frequent sexual intercourse with multiple partners and from smoking, both from a young age. It is associated with the human papillomavirus (HPV) which can also cause genital warts. Another type, called **adenocarcinoma** occurs in women who have never had sexual intercourse.

Ectopic pregnancy occurs when the foetus develops outside the uterus, usually in the fallopian tube. The tube may rupture causing severe bleeding and infection. The pregnancy must be terminated.

Endometriosis is a condition where fragments of endometrial tissue grow outside the uterus and spread to the ovaries or other organs in the pelvic cavity. It is usually found in women of child-bearing age and may cause infertility. It has no known cause, but it is thought that menstrual tissue becomes misrouted into the pelvic cavity instead of being evacuated from the body during menstruation. Symptoms can include very heavy bleeding during menstruation and severe abdominal and back pain.

Fibroids are slow growing benign tumours, which can vary between pea and grapefruit size. They consist of smooth muscle enclosed in fibrous muscle tissue. They may be caused by an abnormal response to the hormone oestrogen. There are often no symptoms, but large fibroids can press against the bladder causing frequent urination, or against the bowel causing constipation or backache.

Ovarian tumours involve different types of cell. Tumours arising from embryonic cells occur in those under 18 and are usually benign cysts. Tumours arising from epithelial cells which cover internal organs may be malignant and spread to other parts through the bloodstream and lymph system. Hormone-secreting cells may produce tumours associated with the abnormal production of oestrogen. This can lead to early sexual development in children, and in adults may cause breast and genital problems

Polycystic ovaries (Stein-Leventhal syndrome) shows several apparently unrelated symptoms, including infrequent menstruation, infertility, acne, obesity and excessive hair growth. It may be caused by abnormal levels of testosterone and luteinizing hormone (LH). Treatments vary according to the symptom.

Vulvitis is an inflammation of the vulva. This can be caused by infections including candidiasis, genital herpes, warts, scabes or pubic lice. Other causes may include allergic reactions to soaps or cosmetics. Infection can also be spread by sexual transmission to the vagina (vulvovaginitis), and cause Pelvic Inflammatory Disease (PID). Complications can cause infertility, peritonitis or meningitis.

REFERENCES

[1] Dewsbury, D.A., (1981), Effects of novelty on copulatory behaviour: the Coolidge effect and related phenomena. *Psychological Bulletin*, **89**, pp. 464-482

[2] Freeman, W., (1973), Sexual behaviour and fertility after frontal lobotomy. *Biological Psychiatry*, **6**, pp. 97 – 104

[3] Alexander, C.J., Sipski M.L., and Findley, T.W., (1993). Sexual activity, desire and satisfaction in males pre- and post-spinal cord injury. *Archives of Sexual Behaviour*, **22**, pp. 217-228

[4] Lundberg, P.O., (1992). Sexual dysfunction in patients with neurological disorders. *Annual Review of Sex Research*, **3**, pp. 121-150

[5] Meredith, M. (2001), Human Vomeronasal Organ Function: A Critical Review of Best and Worst Cases, *Chemical Senses*, **26,4**, Oxford University Press, pp. 433-445

[6] Stern, K., and McClintock, M. K., (1998), Regulation of ovulation by human pheromones, *Nature* **392**, pp. 177-179

[7] Gladue, B.A., (1988), Hormones in relationship to homosexual/bisexual/heterosexual gender orientation. In *Handbook of Sexology*, Vol, 6., *The Pharmacology and Endocrinology of Sexual Function* (ed. J.M.A. Stitson), Elsevier Science, Amsterdam, pp. 388-409

[8] Swaab, D.F., and Hofman, M.A., (1990), An enlarged suprachiasmatic nucleus in homosexual men. *Brain Research*, **537**, pp. 141-148

[9] LeVay, S., (1991) A difference in hypothalamic structure between heterosexual and homosexual men. *Science,* **253**, pp. 1034-1037

[10] Semb, H., "Human embryonic stem cells: origin, properties and applications" *APMIS.* 2005 Nov-Dec; **113** (11-12): pp. 743-50. PMID 16480446

[11] Campbell, K.H.S., McWhir J., Ritchie, W.A., Wilmut I.(1996) "Sheep cloned by nuclear transfer from a cultured cell line" *Nature,***380**, pp. 64-66

CHAPTER 16

SKIN, HAIR and NAILS

What are the functions of the skin?
How is the brain involved with the skin?
What is the skin composed of?
What happens when skin is broken?
What makes skin different colours?
Why do we get goose pimples and shiver?
What causes dry and cracked skin?
Why do we sweat and why do feet smell?
What are fingerprints composed of?
What are blisters?
What is hair composed of?
Why does hair turn grey?
Disorders affecting skin and hair
What are nails made of?

The skin, hair and nails are collectively called the Integumentary System. The name comes from the Latin word meaning to cover or enclose.

What are the functions of the skin?

The skin performs several vital functions for the well-being of the body. It forms a protective layer, which is largely waterproof and is a barrier against invasion from the environment by bacteria. It protects against the ultraviolet rays of the sun and helps to prevent dehydration.

The skin also helps to regulate and control the temperature of the body, keeping it at a fairly consistent level of about 37°C (see sections below on sweating and shivering). The majority of body heat is lost through the skin, although some is also lost when breathing out and eliminating waste material. We also help maintain this optimal temperature by wearing clothes to suit the surroundings. Layers of clothing trap air which helps to retain the heat. This is why in cold weather it is better to wear several layers of lightweight clothes rather than one heavyweight item. The skin also preserves heat produced when metabolic rate is increased. This is caused by organs of the body such as the muscles, which produce heat during exercise, and during the course of digestion, when food is broken down.

How is the brain involved with the skin?

The nervous system reacts to changes to the skin, and the

hypothalamus in the brain reacts to changes in temperature and stimulates the nerves. Increases in temperature cause the tiny capillary blood vessels to dilate through actions of the medulla, bringing more blood to the skin surface. This makes the skin look pink and warm to the touch. When temperature falls the reverse action constricts the capillaries, less blood reaches the skin, which looks white and feels cool. Please see also the chapter on Pain and Touch, which discusses the connection between brain and skin in more detail. There are no nerve receptors in hair shaft or nails, as they are effectively 'dead' layers.

What is the skin composed of?

The skin is made up of three layers: epidermis, dermis and hypodermis. Figure 16.1 illustrates the structure.

Figure 16.1 Structure of the skin

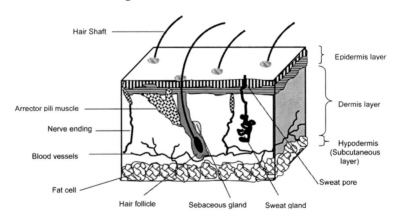

The outer layer is the epidermis, which has five tiers or strata of epithelial (outer covering) tissue. This layer varies in thickness and is deepest on the soles of the feet and palms of the hands. The deepest tier is called the stratum germinativum or basal layer. It has a rich blood supply from the underlying dermis layer, and the cells continually reproduce, pushing old cells upwards. The next tier is called the stratum spinosum or prickle cell layer. The cells change their cuboid shape to spiky – hence the name. As they move towards the outer surface, they form the next tier called the stratum granulosum or granular layer. The cells begin to flatten out and keratinise (harden) as the cytoplasm in the cells is replaced by a fibrous protein called keratin. The process is called desquamation.

The next tier is called the stratum lucidum or clear layer. Enzyme action destroys the melanin pigment tissue, making the cells transparent. The outermost tier is the stratum corneum or horny layer. Here the dead cells called squames, are thin and flat. They flake off as the newer cells develop from the lower tiers. The whole process takes about a month to complete. Figure 16.2 shows the tiers of the epidermis layer.

Figure 16.2 Tiers of the epidermis

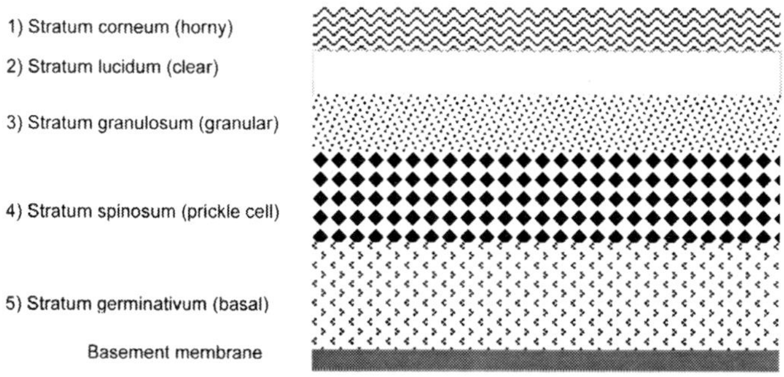

1) Stratum corneum (horny)
2) Stratum lucidum (clear)
3) Stratum granulosum (granular)
4) Stratum spinosum (prickle cell)
5) Stratum germinativum (basal)
Basement membrane

The dermis consists of two layers. The papillary layer connects with the epidermis and has loose connective tissue. It contains sensory touch receptors and tiny blood capillaries which project into and nourish the epidermis. Under this is the reticular layer which has dense connective tissue containing fibres of the proteins collagen and elastin. This provides the skin with strength and elasticity. Reticulin is another protein, and its fibres support hair follicles, sweat and sebaceous glands, arretor pili muscles as well as nerve endings, lymph and blood vessels which are found in this layer.

The hypodermis or subcutaneous layer is made up of two types of tissue. Adipose fatty tissue provides insulation for warmth and acts as a reserve for energy. Areolar tissue supports, cushions and protects through its pliable, loosely weaved structure.

What happens when skin is broken?

The healing process involves three phases: Inflammatory, Proliferative and Mature. The immediate response is the Inflammatory phase when blood platelets accumulate over the wound. An enzyme from the platelets called thromboplastin then begins to coagulate the blood to form a clot. In the process the enzyme becomes thrombin (hence the word 'thrombosis' for blood clot). The blood vessels dilate, causing the surface to become inflamed. Phagocyte cells start to ingest and destroy foreign micro-organisms, and fibroblasts migrate to the area. These cells provide a structural framework for repair work to begin.

The next is the Proliferative phase during which epithelial

cells tightly pack together and proliferate across the wound to form a barrier to prevent further infection.

The final Mature phase is when the epidermis rejoins from upward growth of new skin cells, and the clot above the new tissue becomes a dry scab. New collagen develops to strengthen the area and the scab eventually falls off. The time taken for this process to complete depends on the extent of the damage and the general health of the individual, as well as the environmental conditions.

What makes skin different colours?

Skin colour is genetically determined by the amount of melanin in the skin. Melanin is a dark pigment which comes from the amino acid tyrosine. It is secreted by melanocyte cells in the germinative layer of the epidermis and is absorbed by epithelial cells. The depth of colour depends on the level of secretion. Melanin protects the skin from damage caused by ultraviolet light from the sun, which is why people living in hot climates tend to have darker skins.

The pink colour of so-called white-skinned people depends on the percentage of haemoglobin to blood circulating in the dermis.

The yellowish skin tones of people such as those from China or Japan is caused by disproportionate levels of bile pigments in blood and carotenes in the subcutaneous fat layer of the skin.

Why do we get goose pimples and shiver?

Goose pimples or bumps only occur in hairy mammals. They resemble the plucked skin of a goose which is why they are so called. When we are afraid or cold or very emotionally involved, the hypothalamus in the brain stimulates sympathetic nerve fibres in the arrector pili muscles attached to the hair follicle. This causes the muscles to contract, pulling the hair erect, and raising the surrounding surface skin. In response to cold, air is trapped between the erect hairs, and forms an insulating layer for the body.

In humans, the goose-pimple response to fear or strong emotion is an evolutionary throwback, and no longer serves any useful purpose. When other mammals are afraid, the erect hairs can make them appear larger to an enemy. The same effect can also help secure a mate.

As mentioned before, the hypothalamus controls body temperature by causing effects that raise or lower it back to its optimal level. When the temperature drops, this part or the brain stimulates the nervous system and makes us shiver, causing the skeletal muscles to contract. This action produces energy and heat.

What causes dry and cracked skin?

Sebaceous glands are attached to hair follicles, and secrete an oily material called sebum into hair follicles and skin. The glands are generally found on the scalp and face, armpits and groin. Sebum lubricates and protects the skin and hair. Over-activity can produce oily skin and hair and under-activity produces dry and cracked skin. The sebaceous glands are more active from puberty through middle age. This means that young children and the elderly are more prone to dry skin.

Why do we sweat and why do feet smell?

Sweating is a way of losing excess heat, and is controlled by the hypothalamus and sympathetic nervous system. There are two types of sweat glands: eccrine and apocrine. The sweat is produced in the long coiled part of the gland, which acts as a duct connecting the gland to the skin. The duct opens in a pore when it reaches the skin.

Most of the sweat glands throughout the body in the dermis layer of the skin are eccrine, and are found especially on the soles of the feet, palms of hands and forehead. They are active throughout life. The secretion consists of mainly water with sodium and chloride salts. When the body is resting or the outside temperature is cool, the cells in the straight duct reabsorb most of the sodium and chlorine in the fluid. During periods of excessive sweating such as when taking strenuous exercise or in hot weather, these cells cannot reabsorb the salts. They are carried in the fluid to the skin surface, where it then runs down the skin. When the fluid evaporates, it leaves behind the salts. This is why skin tastes salty. The lack of sodium chloride in the body caused by excessive sweating can trigger dehydration which is why we need to ingest water and salt to restore the balance.

Apocrine glands are mainly found in the armpits and genital areas. They become active at puberty. They open directly onto hair follicles, not pores. The thicker secretion contains fatty acids and proteins, and is slightly yellow in colour. Bacteria on the skin break down the organic compounds and their faeces produce the smell of body odour.

Our feet and hands both produce sweat. The main reason why, unlike our hands, our feet smell, is because the sweat cannot escape into the atmosphere. It is contained in socks and shoes and other footwear, and accumulates. Bacterial activity again leaves a residue which produces the smell.

We also sweat when we are nervous or agitated. Sympathetic nerve action together with adrenaline produced from the adrenal glands, situated on top of the kidneys, activates the sweat glands. This information has led to the production of the polygraph machine which records galvanic skin responses, and is supposed to detect when a person is telling lies.

What are fingerprints composed of?

Fingerprints are the impressions made by friction ridges on the fingers and palms of the hand, or toes and soles of the feet. Every person has a unique pattern of ridges on the outer layer of the skin. The papillary layer of the dermis has projections or papillae which extend into the epidermis and produce a double row of ridges in these areas. They help us to grip objects with the fingers, or keep a grip on the surface below the feet.

What are blisters?

Blisters form when the epidermis separates from the dermis. A pool of lymph and other fluids collects between the layers until new skin is grown from below. It is mostly caused by friction when a hard material is repeatedly rubbed against the skin. Other causes can be from heat or frostbite. Diseases such as chickenpox or impetigo or toxic chemicals also cause blisters to occur. Blood blisters are formed when there is bleeding below the dermis. As the new skin develops, the fluid is reabsorbed.

What is hair composed of?

There are three types of hair found on the human body. The foetus has a covering of fine hair called lanugo. This then disappears, and is replaced shortly before birth by a delicate vellus hair, which is soft and downy, and covers all the body except the lips, palms of the hands and soles of the feet. Coarser terminal hair also grows only on the scalp, eyelashes and eyebrows, and acts as a protection for the head and eyes. After puberty, this type of hair extends to the inside of the ear and nose, underarm and pubic areas, as well as the chest in some males. Figure 16.3 shows the structure of the hair.

Cells in the epidermis move down through the lower layers of the skin and form hair follicles. These cells collect into a bulbous shape, which is why it is called the bulb. At its base is the matrix that is producing new hair. As new cells multiply, they push the old cells upward, which begin to die and keratinise. A root sheath surrounds the cells and the inner cells of the sheath link with hair cells to support the hair inside the follicle. The outer root sheath is formed from the stratum germinativum layer of the dermis. The keratinised layers of dead cells produce the strand of hair. When it protrudes away from the skin it is called the hair shaft.

Figure 16.3 Structure of hair

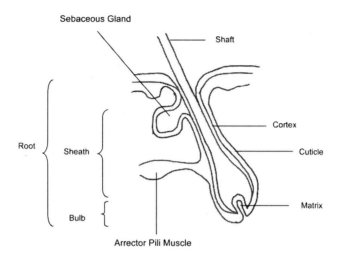

Each hair has three layers: outer root sheath, medulla and inner root sheath. The medulla is in the centre and is only present in thick hair. The cortex contains melanocytes and keratinocytes which determine the colour and strength of the hair. The cuticle is the thicker outer layer with cells that overlap and link with the inner root sheath. Figure 16.4 shows a diagrammatic view of a cross-section of a hair follicle.

There is a three stage growth cycle: anagen, catagen and telogen. The anagen stage is the active phase when cells are multiplying. Hormones activate the outer root sheath to produce new hair from the bulb matrix. Growth stops in the second catagen stage. The outer sheath shrinks and attaches to the root. The final telogen stage is the resting phase, until the hair eventually falls out, leaving room for new hair to develop. Hair on the head grows about 1 centimetre per month, and may last up to six years before falling out. Elsewhere on the body there is an active growth period of only about 40 days, which is why the hairs are shorter.

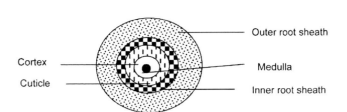

Figure 16.4 Cross-section of hair follicle

Why does hair turn grey?

Grey hairs are part of a natural ageing process. The start of this process varies between individuals and is genetically determined. The cause is loss of melanin. Melanocyte stem cells die and the hair grows with very little melanin to produce the colour. White hair has no pigment at all, the melanin being replaced by tiny bubbles of air.

Disorders affecting skin and hair

Abscess is a local swelling caused by bacteria containing pus.

Acne Roseacea produces a reddish appearance of the nose and cheeks caused when minute capillaries in skin dilate. These are accompanied by pimples and spots.

Acne Vulgaris is caused by infected sebaceous glands resulting in inflammation and spots. These affect the face, neck, back and chest.

Allergies cause the skin to become red, hot and itch.

Alopecia is produced when hair follicles are unable to generate new hairs, resulting in partial or complete baldness. The condition is an autoimmune disease, whereby the body mistakenly identifies hair follicles as foreign tissue.

Asteatosis produces very dry, scaly and itchy skin caused by underactive sebaceous glands

Athletes Foot is a fungal infection, affecting the skin of soles of the feet and between toes, producing itching, flaking and peeling skin.

Boils and **abscesses** are produced when bacteria attacks hair follicles and the surrounding skin. Pus develops and comes to a head. A collection of boils is called a carbuncle.

Bruises result from injury to the skin, causing discoloured skin tissue. Blood leaks from damaged blood vessels into surrounding tissue.

Calluses are caused by abnormal pressure or friction, thickening the skin. They are commonly found on feet or hands.

Cancer of the skin is associated with exposure to ultraviolet radiation from the sun.

Chilbains are painful, reddish areas of the skin on toes and fingers. They are caused by constriction of blood vessels in cold weather.

Chloasma are small patches of dark colour on the skin, produced by abnormal hormone activity, causing stimulation of melanocytes.

Comedone, commonly called blackheads are formed when the pores become blocked with excess dried sebum, dead skin cells, sweat or dirt.

Corns are the central core of thick layers of skin. They are found usually on the toes and soles of feet, caused by frequent friction from footware.

Cysts are small lumps in sebaceous glands. Epidermal cysts appear anywhere on the body. Pilar cysts usually appear on the scalp.

Dandruff or pityriasis is a dry flaky scalp. It is caused when the cell cycle is completed too quickly. As a result, the skin sheds within a week, instead of the more normal monthly cycle.

Exzema or dermatitis is an inflammation of the skin caused when blood vessels dilate. Fluid accumulates and produces swelling, itching and blisters. Weeping skin may become infected and dries to form a crusty scab.

Freckles or ephelides are small, flat, irregular patches of melanin on the face and body. It is caused by an uneven accumulation of melanin in small areas of the skin.

Herpes viruses cause chicken pox and shingles as well as cold sores. Genital herpes is spread by sexual intercourse.

Hives or urticaria is also known as nettle rash. It is an allergic reaction producing raised skin welts and can appear anywhere on the body.

Hypehidrosis is excessive perspiration of the hands, feet and underarms. The cause may be over-activity of the sympathetic nervous system, producing abnormal hormonal activity.

Impetigo is a contagious skin infection. It starts with a red spot that becomes a blister. This eventually breaks down to form a crust spreading across the face, hands and knees.

Moles are dark, raised patches of pigmented skin. They can develop into melanoma, a type of cancer which can be benign or malignant.

Port-wine Stain is a flat patch of dark red staining of an area of skin. This is caused by an abnormally high number of capillary blood vessels. It is a permanent, harmless condition.

Psoriasis is a recurring eruption, producing itchy red patches of scaly skin. It is caused when cells of basal layers of epidermis multiply and move through higher layers of immature cells.

Scabies is a contagious disease caused by infestation of a mite burrowing into skin. It usually affects the areas of skin between the fingers and soles of feet. Can also be found on the wrist.

Warts are small, thick growths on the skin. When found on the feet, they are called veruccas, They are caused by a virus and are contagious.

SKIN, HAIR AND NAILS

What are nails made of?

Human nails are equivalent to claws or hoofs on animals. They develop from cells in the epidermis, which have become hard and keratinised, and protect the fingers and toes. Figure 16.5 shows the structure of the nail.

The root of the nail is embedded under folds of skin called the cuticle. The cells multiply in the root and push the old cells towards the lunula. This is the half-moon (lunar) shape part at the base of the nail. The cells keratinise, but do not scale off (desquamate), and as they get longer, they form the nail plate. Under the plate is the nail bed which adheres to the nail with connective tissue. As the nail grows over the bed, it becomes a free edge.

Figure 16.5. Structure of the nail

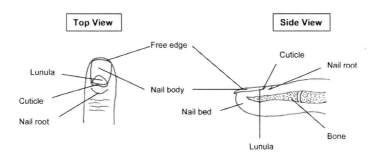

CHAPTER 17

SLEEP and DREAMS

Why do we sleep?
What happens when we sleep?
What is our biological clock?
Can we live without sleep?
Can we learn whilst asleep?
Which brain regions are involved in sleep?
What kind of sleep is caused by general anaesthetics?
Why do we dream?
Why do we have recurring dreams?
Why don't we remember our dreams?
Disorders affecting sleep

Why do we sleep?

There are various theories why we need to sleep every day. The major ones are that the body needs to rest, that during sleep the body restores and repairs itself and that it is an evolutionary process to protect us against predators.

Does the body rest while we are asleep? We do undoubtedly use less energy. During certain phases of sleep the muscles are in a very limp and relaxed state. The heart rate, blood pressure, respiration and body temperature are all reduced from normal levels when awake. However there is a period of sleep where the brain is very active and can affect all these processes. Also we feel the need to sleep each night, whether we have been physically or mentally very active during the day or have spent the day lounging around doing very little. So rest itself does not seem to be the only reason for sleep.

Does the body restore and repair itself whilst we sleep? During slow wave sleep (see next section), the Growth Hormone Somatotropin(GH) is secreted from the pituitary gland in the brain. GH stimulates amino acids to synthesise proteins that restore the cells of body tissues. But these amino acids are only available about five hours after a meal. This is before they have been naturally incorporated into other body processes, such as making their own proteins, or being converted into fats for storage in adipose tissue,[1] so the window of opportunity for GH is fairly limited. However, GH production has its most effective peak soon after the onset of sleep. This is because corticosteroid hormone secretions, which break down proteins, are diminished.

The other theory commonly discussed, is that all vertebrates sleep in a usually sheltered place and blend in with their environment. This protects them from predators and helps to prevent injuries during the night, when visibility is poor. It is also more difficult to find food in the dark. In our early history, these factors may have been very important to humans too, and sleep is an evolutionary hangover that has persisted, as it still provides benefits for optimal performance during the day. Some birds and animals have developed extremely acute senses of sight and smell and hearing and have reversed the process by sleeping during the day and hunting at night under cover of darkness. They make use of the vacuum left by the majority of sleeping animals.

What happens when we sleep?

There are two types of sleep: non-REM, also known as slow wave (SWS) and rapid eye movement sleep (REM). Non-REM sleep has four distinct stages. These cycles of sleep repeat throughout the night four or five times at approximately 100 minute intervals. At the beginning of the night there is more stage 3 and 4 SWS and none at the end of the night which consists of REM sleep.

When we are awake there are two basic patterns of brain activity as measured by an electroencephalograph (EEG) recording: alpha and beta waves. Alpha waves occur when we are relaxed and calm, with regular, medium frequency waves of about 8-13 Hz (cycles per second). They show a synchronised neural activity in the cerebral cortex. Beta waves occur when we are alert and thinking, and have irregular frequency waves in the range of 14-30 Hz. This indicates a higher level of variety of activity.

Brain activity changes when we are asleep. During stages 1 and 2, irregular theta waves of about 4-8 Hz occur with irregular bursts of activity of 12-14 Hz called spindles, as we begin the deeply relaxed condition of early sleep. We then enter into a profound, dreamless sleep at stages 3 and 4 and slow delta waves appear of 1-3 Hz. This is followed by a highly active period of REM sleep with beta waves over 12 Hz. This is the phase when dreams are experienced. Figure 17.1 illustrates the different types of wave pattern.

During stage 1 sleep the heart rate slows and muscles relax. The eyelids slowly open and close and eyes roll slightly. We can notice this when watching other people napping. This stage lasts up to ten minutes. In stage 2 there are also sudden peaks and troughs of activity from waves called K complexes which follow spindles. If awoken at this stage, even after strong snoring, people do not believe they have been asleep. This stage lasts about 15 minutes.

Figure 17.1 Wave patterns

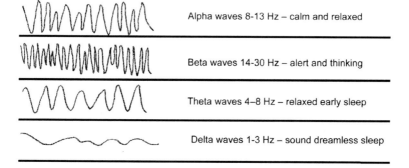

Stage 3 sleep alternates between theta and delta waves and lasts a few minutes before stage 4 sleep begins. If awoken at this stage, a person will feel sluggish and bewildered. These two stages occur mainly during the first half of the night. After about 45 minutes there is a complete change of brain activity similar to that when awake, as the REM stage starts. Under closed lids the eyes move rapidly from side to side as we start to follow the dream. Most of the muscles of the body (apart from the eye muscles) become semi-paralysed. Although we can sleep through a variety of loud noises, such as thunder during this stage, we can be easily awoken by sounds that have some special meaning for us, such as someone calling our name or the cry of a child. If we are awoken at this time, we are alert and can report the dream we have been having. REM sleep lasts about 25 minutes and during an eight-hour sleep there will be four or five further periods alternating with other non-REM sleep.[2] The first REM period lasts about 5-10 minutes, and the last REM period can extend to 40 minutes just before we wake up.[3]

During REM sleep, when we are dreaming, the brain is very active and other consequential physiological changes occur, such as blood flow increases to the brain and more oxygen is used. During dreams with a highly sexual content, the penis can become erect and semen will be ejaculated – the so-called 'wet dreams' of adolescent boys.

What is our biological clock?

Our behaviour is regulated by a range of biological timing mechanisms and our sleep-wake routines are controlled by a circadian rhythm (from the Latin words for 'about a day') which actually lasts for 25 hours, not the 24 hours we usually associate with the length of a day. The circadian clock starts to operate a few weeks from birth. Young children sleep for about 10 hours a night, and adults about 7-8 hours. It is thought that children need to sleep longer to benefit their developing brains. Artificial light during dark periods extends our waking time, and keeping the curtains or blinds drawn over the windows in early morning extends our time sleeping. Under natural conditions, we would probably go to sleep soon after dark and wake up just after dawn.

The body's daily biological clock is situated in a part of the hypothalamus in the brain called the suprachiasmatic nucleus (SCN), just above the optic chiasm where the optic nerves from each eye meet and cross. Light travels to photoreceptors in the retina at the back of the eye which send information about light and dark through the optic nerve directly to the SCN.

There are several genes that control the effects of light and dark on our bodies. The SCN makes two proteins called clock and cycle. These bind together to the DNA of the cell which transcribes two genes called period (per) and cryptochrome (cry). These bind with another protein called tau and together they inhibit more production of clock and cycle, making them unable to process further transcriptions. The per/cry/tau protein complex gradually breaks down and after 24 hours, the cycle starts again.[4]

The SCN also sends signals through other hypothalamic nuclei to the spinal cord, and influences the activity of the pineal gland in the centre of the brain. This gland secretes melatonin during the night into the blood stream. One effect of melatonin is to induce the feeling of drowsiness. Our ability to sleep depends partly on the level of melatonin secretion. The SCN also synchronises physiological processes related to body temperature, urine production and blood pressure levels.

Can we live without sleep?

Nearly all of us need to sleep every day. Sleep patterns amongst the elderly seem to vary, with less sleep at night and more catnaps during the day. Lack of sleep does not appear to adversely affect physical activity. However, being deprived of sleep for a considerable period can lead to unreal perceptions, such as hallucinations. There was a study in the U.S. in 1965 of a boy of 17 who decided to beat the world record for staying awake to get himself into the Guinness Book of Records. He managed to stay awake for 11 days. He then slept for 15 hours, after which he seemed fit and alert. In the subsequent days he never made up his lost hours of sleep, although the patterns of sleep were different, with over half of the later stages of SWS and also REM sleep being made up.[5]

Can we learn whilst asleep?

There have been over the years various products on the market which imply we can learn more efficiently – and without the usual interruptions, whilst we are asleep. This is a particularly appealing suggestion to students just before an exam or speakers trying to memorise an important speech for a conference. Some people have tried to learn a foreign language in their sleep. Regrettably, there is no evidence that it works. It seems that we need to pay attention in order to remember – and that can only be done when we are awake.

However, there is some evidence from studies, that recently acquired learning can be consolidated during sleep. If a list of nonsense words is learned, followed by a period of sleep in the next 24 hours, more of the words will be remembered, than if there had been no sleep within 8 hours from learning.[6] Another experiment indicated that the same brain areas involved in recently learned material are also activated during REM sleep.[7]

Which brain regions are involved in sleep?

There are four areas of the brain that interact and control sleeping and waking cycles. They involve the hypothalamus, the base of the forebrain, the pons and the brainstem. Figure 17.2 shows the approximate locations of these organs in the brain.

The hypothalamus contains neurons that produce the peptide protein receptor called hypocretin, which helps us to stay

Figure 17.2 Brain regions involved in sleep

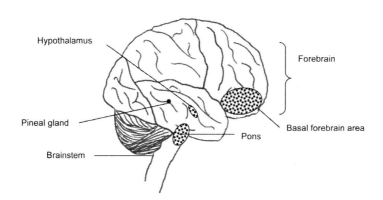

awake. The neuron axons spread through the other three areas to stimulate or inhibit SWS or REM sleep. A deficiency of hypocretin causes the sleep disorder called narcolepsy.

The basal forebrain region generates SWS sleep. Neurons are sensitive to temperature changes. This may be why we feel sleepy when we are hot. Sleep may be a way of keeping the brain temperature at optimal levels. Neurons in the basal forebrain also inhibit the release of a chemical called acetycholine, which keeps us alert and awake.

The pons produces REM sleep. If artificially stimulated chemically or electrically, the result is further periods of this type of sleep. It is also the part that causes the muscles to become immobile by sending messages through the brain and spinal cord to prevent muscle stimulation. This prevents us from physically acting out our dreams. We become watchers only, similar to looking at a television or stage show.

The brainstem contains an area in its central core called the reticular formation. It is an interacting network of cell nuclei which influences many other regions of the brain, and affects sleep patterns. Figure 17.3 shows its location. It effectively wakes up the forebrain from SWS sleep using an arrangement called the ascending reticular activating system (ARAS). The ARAS receives sensory information from the brain and spinal cord which stimulates it to trigger wakefulness. Other nuclei in the brainstem have an opposite effect.

There are three groups of neuron networks in different parts of the brainstem that operate separate systems and have distinct functions: the serotinergic, the noradrenergic and the cholinergic.

Figure 17.3 Reticular formation

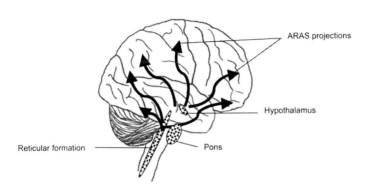

Serotonin is a neurotransmitter that induces sleep. Studies have shown that blocking the receptors that produce serotonin causes insomnia.[8]

Noradrenaline secretion causes wakefulness. The neurons project signals to the cortex which makes us feel attentive and aware of our surroundings.

The cholinergic system arises from the secretion of the neurotransmitter acetylcholine (ACh), which triggers a desynchronised pattern of wave activity seen during REM sleep. The neural circuits activate regions of the brain involved in eye movements and inhibit muscle action during this phase of sleep. They also project to the amygdala, which is a part of the brain involved with emotions. This could account for the emotional content of dreams.[9]

What kind of sleep is caused by general anaesthetics?

The unconscious state induced by general anaesthetic drugs is not the same as sleep. Whilst asleep, several brain regions remain active to varying degrees throughout the night. The unconscious state reduces the flow of sodium molecules into nerve cells which prevents neurotransmission. The brain therefore is receiving very little information. Anaesthetic drugs increase the sensitivity of receptors, which secrete a neurotransmitter called GABA in the forebrain. This acts directly on the central nervous system to inhibit further transmissions which would otherwise keep us awake. It maintains a slow wave sleep. The drugs inhibit excitatory neuro-transmitters such as glutamate and acetylcholine, whilst at the same time boosting the action of inhibitory neurotransmitters such as GABA and glycine. The action of the drugs is temporary, and the dose depends on the amount of time needed for a patient to remain unconscious.

Why do we dream?

There have been many theories about why we dream, both physiological and psychological (if you accept the differential).

In his famous book on the interpretation of dreams, Sigmund Freud suggested that our dreams are fulfilling subconscious wishes.[10] According to his theory, there is a conflict between how we would like to behave and what our conscience (or society) deems proper. This may be why the content of our dreams is often very aggressive or erotic. Dreams overcome this tension by using symbols to replace our wishes, which become displaced representations. That is why our dreams often appear unreal and weird.

The Activation-Synthesis hypothesis[11] in the late 1970's,

proposed that the brainstem produced signals, which were interpreted by the cortex, and activated internally generated information to produce dreams and rapid eye movements.

This was followed by a Continual-Activation theory of dreaming,[12] which proposed that the function of dreaming is to transfer data acquired during the day from a temporary to a long-term memory store. The information is collated and updated, and any unwanted memories are deleted. This happens during the Type 1 dream which is non-REM sleep. The transfer triggers a continual activation mechanism in the brain which randomly retrieves information to produce Type 2 REM dreaming. This means that dreaming occurs during both stages of sleep.

Another recent theory is that dreaming is a kind of unfocused method of dissembling anxieties that occur throughout the day, and strengthening neural connections. This will help us to cope with the stressful situation. People who are recovering from a traumatic incident, or who are worried about an important forthcoming event, such as a wedding or moving house, tend to have more REM sleep and more vivid dreams.

Why do we have recurring dreams?

Most of us at some time have had a dream that has a common theme, such as trying to climb over a steep hill, falling from a great height or running in a public place stark naked. People who have experienced a traumatic event can relive the experience many times in their dreams. Unfortunately, there is no definitive answer to this question. However these dreams generally involve stress and anxiety in our daily lives.

Attempting to reach the top of the hill may relate to trying to reach a goal. Falling indicates insecurity and loss of control of a situation. Being naked may be the inability to hide something that we may prefer not to disclose.

The peculiar nature of dreams suggests that the areas of the brain involved in memory are transmitting via specific pathways. This enables us to come to terms with or resolve the situation leading to our anxieties. Although there are interpretations for these types of dreams, the form they take has not yet received a satisfactory neurological explanation.

Why don't we usually remember our dreams?

Freud suggested that we forget our dreams, because they contain repressed desires that we do not want to remember whilst awake. But the general consensus of opinion is that dreams are not important enough for us to remember. Most dreams are forgotten within minutes of waking up, although some event during the day can trigger a partial recall.

Another explanation is that images or ideas need to be reinforced through repetition and association for the information to be retained in long-term memory.

Disorders affecting sleep

Insomnia is inadequate amounts of sleep throughout the night for a prolonged period. Sufferers may have difficulty falling asleep or staying asleep. EEG recordings have shown however, that the total amount of sleep is often underestimated. It affects people under stress, under the influence of drugs or excessive levels of alcohol or in pain from unrelated disorders. One theory of the cause is overactive noradrenaline receptors in the brain stem.[13]

Narcolepsy is suddenly falling asleep for a few minutes only, at inappropriate times of the day. There are also related conditions. Cataplexic patients will suddenly lose voluntary muscle tone and consequently fall down whilst remaining completely conscious. This also lasts for only a few minutes. At the beginning of sleep or at the end of the night, two other temporary, related conditions can also occur: total muscular paralysis or vivid hallucinations. The conditions appear to be caused by abnormal neural activity in the brain, causing a loss of the first four sleep phases. There may be a genetic abnormality, as the conditions tend to run in families.

Nightmares are frightening dreams that occur during REM sleep and will often cause us to wake at a particularly scary part of the dream. They are easily recalled because of the vivid nature of the images. It is thought the brain causes us to wake to alleviate the stress.

Night terrors in contrast occur during slow-wave sleep and are more usual in children. The content cannot be recalled, only the feeling of dread and in some cases suffocation caused by an apparent inability to breathe. Bed-wetting and sleep-walking also takes place during this phase of sleep. They appear to be triggered by daytime stress, and may also be genetically inherited.

Restless leg syndrome (RLS) is an irresistible urge to move the legs while at rest. The feeling is difficult to describe, but moving alleviates it. There is no definitive cause, but various suggestions indicate symptoms are associated with the use of some medications, pregnancy, varicose veins and unusual neural impulses from the brain. Research has also implicated dopamine and iron deficiencies.[14]

Sleep apnea occurs when the tongue and soft palate in the mouth lose muscle tone, causing an obstruction in the airway. Breathing temporarily stops. Lack of oxygen and a build up of carbon dioxide in the brain forces the sleeper awake and gasp for breath. This can happen on and off throughout the night, disrupting the normal sleep phases. Snorers and the obese are particularly susceptible to the condition.

Sleepwalking (Somnambulism) occurs when sleepers arise from bed and carry out functions which have happened during the day. It takes place in the early part of the sleep period during SWS sleep, when the sleeper is in a low conscious state. Activities can include getting out of bed and moving around the bedroom or other rooms, or even going outside the home. Sleepwalkers have been known to take a drive or cook during these spells. Occasionally, adult sleepwalkers can make violent gestures or actually commit violence. The eyes are open, but after the event, the sleeper is unaware of the occurrence.

The condition is most prevalent in children and may relate to an immature central nervous system. It usually disappears after adolescence. The National Sleep Foundation[15] estimates that up to 15% of the population can have periods of sleepwalking.

It is a common misconception that it may be dangerous to

waken sleepwalkers. They will be disorientated and confused, but are not harmed. Another misconception is that they avoid injury whilst sleepwalking. They often do trip and have been known to fall out of windows.

There is no definitive cause, but in children anxiety seems to have a role. In adults the causes may be associated with drugs or alcohol or a brain disorder. No treatment is necessary in most cases.

REFERENCES

[1] Carlson, N.R., (1994), *Physiology of Behaviour*, Fifth Edition, Allyn and Bacon, p. 264

[2] Carlson, N.R., (1994) *Physiology of Behaviour*, op. cit., p. 256

[3] Rosenzweig, M.R., Breedlove, S.M., Watson, N.V., ((2005), *Biological Psychology*, Fourth Edition, Sinauer Associates, Inc., p. 436

[4] Rosenzweig, M.R., Breedlove, S.M., Watson, N.V., (2005), *Biological Psychology*, op. cit., p. 430

[5] Gulevich, G., Dement., W.C., and Johnson, L., (1966), "Psychiatric and EEG observations on a case of prolonged (264 hours) wakefulness", *Archives of General Psychiatry*, **15.**, pp. 29-35

[6] Gregory,R.L. (ed) (1987), *The Oxford Companion to The Mind*, Oxford University Press, p. 719

[7] Maquet, P., Laureys, S., Peigneux, P., Fuchs, S., et. al., (2000), "Experience-dependent changes in cerebral activation during human REM sleep", *Nature Neuroscience*,**3.**, pp. 831-836

[8] Jouvet, M., (1969), "Biogenic Amines and the States of Sleep", *Science* **163** (862), pp.32-41

[9] Toates, F., (2001), *Biological Psychology*, Prentice Hall, p. 492

[10] Freud, S., (1900), *The Interpretation of Dreams*, 3rd Edn. Translated by A.A. Brill

[11] Hobson, J.A., and McCarley, R., (1977), "The brain as a dream state generator: An activation-synthesis hypothesis of the dream process". *American Journal of Psychiatry,* **134**, pp. 1335-1348

[12] Zhang, J., (2004), "Memory Processes and the function of sleep". *Journal of Theoretics, Vo. 6.6*

[13] Hobson, J.A., (1999), "Sleep and dreaming" in *Fundamental Neuroscience* (eds. M.J. Zigmond, F.E. Bloorm, S.C. Landis, J.I. Roberts and L.R. Squire), Academic Press, pp.1207-1227

[14] Connor, J., Boyor, P., Menzies, S., Cellinger, B., Allen, ~R., Ondo, W., and Earley, C., (2003) "Neuropathalogical examination suggests impaired brain iron acquisition in restless leg syndrome", *Neurology* **61** (3), pp. 304-9

[15] www.sleepfoundation.org.

CHAPTER 18

TASTE and SMELL

How do we discriminate between different tastes?
How does the brain receive taste information?
Why do we lose our sense of taste when we have a cold?
Are you a supertaster?
How do we discriminate between different smells?
How does the brain receive smell information?

How do we discriminate between different tastes?

Taste is known collectively as the gustatory system. There are four basic types of taste: sweet, salty, sour and bitter. Even very young babies can discriminate between them, obviously enjoying sweet, and spitting out bitter tastes. Sweet tastes are usually preferred, and generally relate to food which has a high calorific value. This is probably because a high calorie intake is beneficial for the highly energetic lifestyle many humans pursued in the past. Interestingly, most animals are able to respond to the same four tastes, except members of the cat family of all kinds including lions, tigers and domestic cats. Studies have shown that they have a genetic mutation that renders them unable to respond to sweet tastes.[1] So there's no point in giving your cat a chocolate treat!

The body needs a certain amount of sodium chloride (salt) and acids in order to carry out its normal digestive functions, so salty and sour foods are also necessary in limited quantities. When we are wounded, there is a loss of sodium chloride in the blood. After vigorous exercise or in a very hot atmosphere, there are also losses in sweat through the skin. At these times we tend to prefer to eat more salty and sour foods when they are available, in order to restore the body to normal levels. Bitter tastes usually relate to foods that are poisonous or stale, and we have an inherent dislike of these tastes, based on the experiences in our evolutionary past. Many plants produce protective alkaloids that taste bitter and prevent them from being eaten. We try to avoid tastes that are associated with periods of ill health.

The tongue has taste receptor cells in little bumps all over its surface. These are called papillae. Each papilla has at least one taste bud, with about 100 – 150 receptor cells. Taste buds are also found on the soft palate, pharynx and larynx in the throat. A total of about 10,000 taste buds are in the mouth and throat. There are three types of taste papillae: fungiform, foliate and circumvallate. Fungiform are slightly mushroom-shaped and are found mainly at the tip and along the sides of the tongue. Foliate papillae have ridges and grooves and are located along the edges of the back of the tongue. Circumvallate papillae are found in a semi-circle at the back of the tongue. Figure 18.1 shows details of the tongue, papillae and taste bud.

Figure 18.1 Tongue, papillae and taste bud

| TONGUE | PAPILLAE TYPES | TASTE BUD |

Circumvallate

Foliate

Fungiform

Taste pore

Taste cell

Nerve fibres

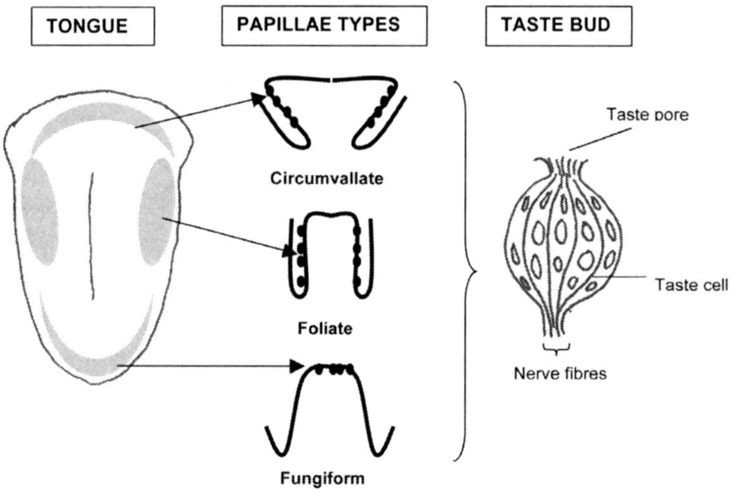

Taste buds have some receptor cells that respond to taste, while others react to pain or touch. Each bud is arranged into groups of receptor cells like the segments of an orange. The cells live for about 12 days, and each bud contains cells at various stages of development and replacement. Minute hairs project through the pores at the tongue surface to mix saliva with chemicals in the food. Tastes can only be analysed in solution, so food entering the mouth triggers the production of saliva. The smell of food can also act as a conditioned reflex to produce saliva and gastric juices (see also chapter on Learning and Memory).

The tip of the tongue is most sensitive to sweet and salty tastes, the sides to sour tastes and the back parts of the mouth and throat are most sensitive to bitter tastes. However, all areas of the tongue are able to distinguish each type of taste. Some scientists believe they have found a fifth taste, which is called umami. The taste is savoury and is found in meats, cheeses and other protein foods. Receptors respond particularly to monosodium glutamate which is added to many processed foods for extra flavour. However, there is some dispute about the hypothesis, and more evidence is required for this extra sense to be officially recognised. Flavours are a mixture of many different chemicals contained in the food we eat, and it may be that the savoury taste is a blend of the four basic tastes.

How does the brain receive taste information?

The substances we place in the mouth are dissolved in saliva. Each type of receptor binds with a different chemical to produce the various tastes. Salty receptors react with sodium ions and sour cells react with hydrogen ions to depolarise ion channels and produce signals for the release of neurotransmitters. Sweet and bitter substances bind with receptors that activate enzymes which will cause the release of neurotransmitters. Figure 18.2 illustrates the main pathways from the tongue to the brain.

The signals pass through three cranial nerves. Cranial nerve VII is the facial nerve and branches travel over two thirds of the tongue from the front. Cranial nerve IX is called the glossopharyngeal nerve and covers the back third of the tongue. The vagus nerve is cranial nerve X and branches over areas in the throat.

Figure 18.2 Main neural pathways of gustatory system

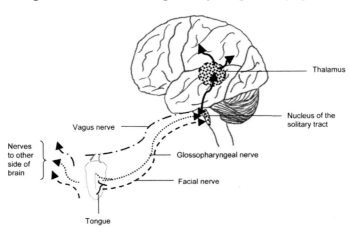

Impulses travel along the nerves to a structure called the nucleus of the solitary tract. This is located along the length of the medulla in the brainstem. Here, taste sensitive neurons relay information to the thalamus and then on to taste regions in the somatosensory cortex.

Why do we lose our sense of taste when we have a cold?

The senses of taste and smell are closely related. When we

have a cold, the nasal passages are inflamed, preventing smells from reaching the sense receptors at the back of the nose. This also has a side effect of lessening the perception of taste. Try an experiment. With eyes closed and nostrils pinched closed with fingers, it is almost impossible to tell the difference in taste between a slice of apple, turnip or even an onion. We have a reduced sense of taste when the mouth is dry, because, as mentioned previously, chemicals must be dissolved in saliva solution to activate the receptors.

Are you a supertaster?

There are some people who are very sensitive to certain tastes, and particularly dislike salty or bitter tastes and have a strong preference for sweet tastes. They probably have more than usual taste buds. A little experiment can test this out. Place some blue or green food colouring on the tip of the tongue with a cotton bud. Cut out a small piece of paper around a punched hole, or use a reinforcing ring (that normally goes over the hole in punched paper), and place it on the coloured part of the tongue. Count the little pink dots inside the ring, using a magnifying glass. These are fungiform papillae. If there are more than 30, then you are a supertaster. Conversely, if there are less than 15 you have a comparatively poor sense of taste overall.

How do we discriminate between different smells?

The sense of smell is also known as the olfactory system. Some smells are pleasant, such as the smell of food cooking when we are hungry, or the perfume of flowers. Others are to be avoided, such as rotting food or one that is connected to danger, such as the smell of gas. The state of the body also defines our preferences. For example, some women in the early stages of pregnancy may be repulsed by certain smells which can make them vomit.

Molecules of chemicals are inhaled through the nostrils, and the air is guided through three long narrow curled passages of bone called turbinates to meet receptor cells at the back of the nose. The structure of the turbinates effectively increases the surface area inside the nose, to enable the smell molecules to reach the maximum number of receptors. The olfactory region is roughly 2.5 square centimetres of the epithelium at each nasal passage. The epithelium consists of cell tissue and mucous membrane. There are a total of 40 – 50 million sensory receptor cells in these two areas. Unlike the sense of smell, receptors are not grouped into categories. There are hundreds of different types of receptor and each one relates to a specific chemical molecule. The brain is able to interpret the information from many different receptors. We are also able to identify different combinations of smell simultaneously. For example, as we walk along the street, we can detect the smell of fish when we pass a Fishmonger's shop, as well as exhaust fumes from an overtaking car and smoke from a nearby bonfire. Each type of smell that we identify is made up of many different chemical molecules.

Receptor cells only live for about 40 days and are constantly replaced. Each contains tiny hair-like processes that extend into the mucous membrane. Chemical molecules dissolve in the mucous and stimulate the receptors to begin the transmission towards the brain. Sniffing also helps to quickly draw in air towards the olfactory region.

How does the brain receive smell information?

The mucous membrane contains proteins that interact with chemicals in the air, and trigger the bipolar receptor neurons to send signals along its axons towards the brain. These nerve endings travel through the cribriform bone, and converge to form round structures called olfactory glomeruli in the olfactory bulb. Each one has an accumulation of about 1000 axons. Signals are then passed from the glomeruli to mitral cells, whose axons form the olfactory tract. Further signals travel through the tract to the piriform cortex, the amygdala and the hypothalamus. Figure 18.3 shows the brain regions involved in the olfactory system.

Figure 18.3 Brain regions of Olfactory System.

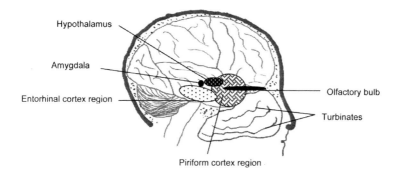

Another important region of the brain also affected is called the entorhinal cortex, which is a memory centre. It is one of the first areas to be affected in Alzheimer's disease, and may be why patients often report a loss of smell, caused by deterioration of this part of the brain. Figure 18.4 illustrates the organisation of the structure of the olfactory system.

Figure 18.4 Structure of Olfactory System

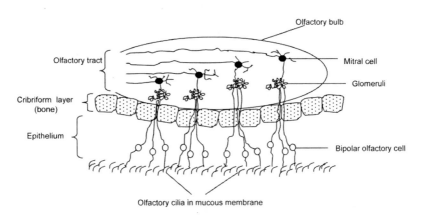

Information relating to a smell may be stored in the entorhinal cortex, and may also be the reason why a smell can remind us of an occasion or experience that occurred a long time ago.

REFERENCES

[1] Li, X., Li, W., Wang, H., Bayley, D.L., Cao, J., Reed, D.R., Bachmanor, A.A., Huang, L., Legrand-Defoetin, V., Beauchamp, G.K., and Brand, J.G., (2006), "Cats lack a sweet taste receptor", Journal of Nutrition 136 (7 Suppl) pp. 1932S-1934S

CHAPTER 19

VISION

How do patterns of light become images in the brain?
What causes light to become coloured?
What is colour-blindness?
Can we trust what we see – illusion and reality?
Why do we need to make eye-contact?
Does eating carrots improve eyesight?
Disorders affecting the eye

How do patterns of light become images in the brain?

Figure 19.1 Cross-section of the human eye

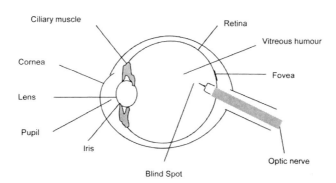

Figure 19.1 shows the main parts of the eye involved in vision. Accurate optical images are required before we can see the shape of objects. When light travels in a straight line, it meets a change in the density of the medium, which causes the light rays to bend. This is called refraction. A simple experiment will show refraction at work. Place a straw or pencil in a glass of water at an angle, and you will notice the angle will appear to change as it enters the water. The cornea is a layer of transparent connective tissue which allows the light to pass through. The lens changes shape to allow light impulses to focus on the retina. The movement of the eyes is controlled by three pairs of extraocular muscles, which extend from outside the eyeball to the bony socket of the eye. Eye movements can be smooth or jerky (saccadic).

The pupil is a hole through which the light impulses enter. Its diameter is altered by dilating or constricting the muscles of the iris, which is the coloured part of the eye. In bright light, the pupil contracts quickly admitting about one-sixteenth as much light as when the illumination is dim.[1] Light coming from the right half of the visual field arrives at the left half of each retina, and vice versa.

The focus of the image is adjusted by changes that occur in the shape of the lens, and are controlled by the ciliary muscles at each side of the eye. When the muscles contract or relax, the lens focuses near or far objects so that a sharp image is formed on the retina. The patterns of light pass through the vitreous humour, which is a clear liquid and fills the space between the lens and the retina.

Processing the image begins in the retina, which consists of several types of layered cells: photoreceptors, bipolar cells, ganglion cells, horizontal cells and amacrine cells (see Figure 19.2). The retinal image is inverted and reversed right to left compared to the visual field. There are two types of photoreceptor cells called rods and cones. These convert patterns of light into a neural image. They release neurotransmitter molecules that transmit information through their synaptic connections with the bipolar cells. These connect with ganglion cells whose axons form the optic nerve. This carries information to the brain, by conducting electrical impulses called action potentials along the visual pathways. The photoreceptor cells connect laterally with the horizontal cells. Finally, amacrine cells also connect laterally with the bipolar and ganglion cells. All of these different types of cell affect each other by releasing neurotransmitters. Small electrical impulses produce the changes. The amacrine cells inhibit the release of the transmitters, where appropriate.

Figure 19.2 Structure of retina

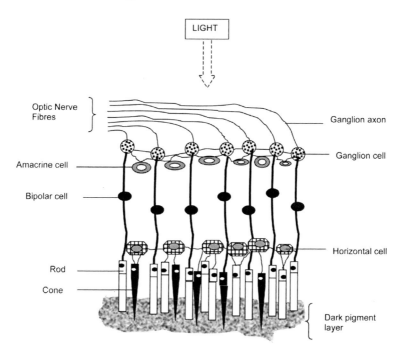

The human eye consists of about 100 million rods and 4 million cones, with about 1 million ganglion cells transmitting information to the brain.[2] The rods react in dim light and do not respond to different wavelengths. This is why at night most objects are seen in shades of grey. The cones respond to colour and provide detailed neural information. The photoreceptors can alter their responses to differing levels of illumination. The pigment in the photoreceptor is rhodopsin, which has two components: retinal and opsin. These are split by light and slowly recombine. The rate at which they recombine determines how much photopigment is available to react to a light stimulus. This can be tested when going from bright daylight into a dark room. It takes several seconds before there is enough rhodopsin available to restore your vision in the dark. The fovea is a small depression or dip in the centre of the retina (see Figure 19.1) which has a dense concentration of cones. This responds to the fine detail of an image.

The periphery of the retina has an abundance of rods and a pooling of outputs. The result is a high sensitivity, with the capacity to detect weak lights. This can be tested by carrying out a little experiment. Stare in a focused way at a very distant star in the night sky. It will probably disappear after a short time. Staring hard corresponds to bringing the image to focus at the fovea, whose cells have a low capacity to integrate weak light over a large area. Look to one side of the star and it should reappear. This corresponds to the image falling on the periphery of the retina.

Each ganglion cell receives information from several photoreceptors, some in the fovea but mostly in the periphery. The receptive field of most ganglion cells has two concentric circles. These are excited when light falls in one area and become inhibited when it falls in the other. This enables the nervous system to detect levels of brightness. ON cells are excited by light in the centre and OFF cells are excited by light in the surround. ON cells distinguish light objects against dark backgrounds, whereas OFF cells distinguish dark objects against light backgrounds.

Axons are long extensions from a nerve cell that transmit nerve impulses from the body of the cell to other neurons. Axons from the retinal ganglion cells carry information from the photoreceptor cells, and travel along the optic nerve towards the brain via various pathways.

The neuronal pathways from the left half of the eye go to the left half of the brain and vice versa (see Figure 19.3). There is a crossover point halfway along the pathway at the optic chiasma from each eye to the other side.

Axons from the nasal halves of each eye cross to the optic tracts on the opposite side. The rest stay on the same side. Therefore each cerebral hemisphere gets input from both eyes. About 90% of axons in the optic tract terminate in the lateral geniculate nucleus (LGN), which is in part of the area of the brain called the thalamus. The remaining 10% terminate in the superior colliculus (SC), which is in part of the midbrain (the inferior colliculus receives information relating to hearing).

Figure 19.3 Visual Pathways

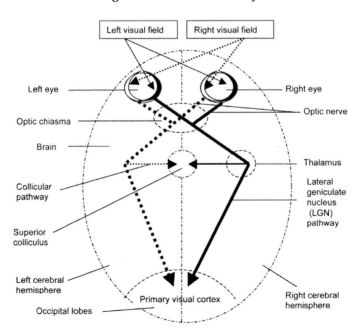

The collicular pathway to the SC bypasses the cortex, and is a fast route to processing information. But this only gives a rough idea of the location of the image. It is especially sensitive to stimuli on the periphery of the retina, for example as seen out of the corner of the eye. A sudden movement would startle and hold our attention, and the colliculus directs a quick, reflex eye movement towards the stimulus. It is thought to be the older of the two main visual routes, and is found in most animals. It is especially useful when looking for predators or searching for prey. It also enables us to perform accurate movements such as hitting a ball with a bat or catching a ball.

The LGN pathway travels from the thalamus to the primary visual cortex. It supplies detailed information relating to the identification and analysis of the image received from the visual field. It is considered to be a newer evolutionary development. The pathway divides into the ventral stream at the bottom of the brain and the dorsal stream at the top of the brain, and consists of six layers. Layers 1 and 2 on the ventral side are called the magnocellular layers with large cells (see Figure 19.4). The cell axons follow a pathway through the primary (striate) and association (extrastriate) visual cortex and on to the posterior parietal cortex. The remaining four layers on the dorsal side are called the parvocellular layers and have small cells. Their axons also travel through different areas in the visual cortex and lead on to the inferotemporal cortex. (see Figure 19.5).

The ventral stream ends in the inferotemporal cortex and in evolutionary terms is a more recent development. It processes our conscious visual experiences and is involved in planning and decision-making.[3] It also links into the parts of the brain involved in learning and memory, such as knowing how to ride a bicycle or tie shoelaces. It is sometimes called the 'what' pathway (see also chapter on Learning and Memory).

Figure 19.4 Pathways to the LGN

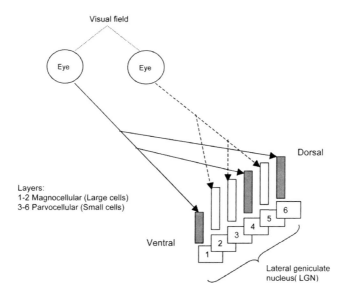

Figure 19.5 Visual processing streams

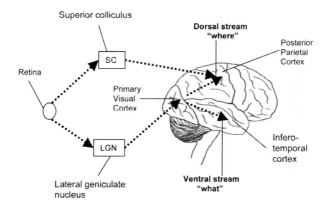

The dorsal stream ends in the posterior parietal cortex and is known as the 'where' pathway. It guides our immediate movements such as reaching out and picking up a pen. We perform these actions quickly and unconsciously.

The visual cortex is divided into several areas along the pathways. The primary visual cortex (V1 – striate) is in the occipital lobe and is specialised to process information relating to moving objects and pattern recognition. It receives its information from the six layers of the LGN. It transmits information to all other areas.

Area V2 receives information from V1 and transmits to V3, V4 and V5. It also has a feedback mechanism where it sends back information to V1. The cells are tuned to orientation, spatial frequency and colour. Spatial frequency is the analysis of the speed with which a property of a stimulus changes in space. Stimuli with fine detail or sharp edges have a high spatial frequency and those whose properties change slowly have a low spatial frequency.

V3 has connections from V1 and V2, and projects strongly with the inferotemporal cortex. It is thought that it may also contain complete visual representation.

V4 in the extra-striate or association visual cortex receives information from V2 and sends information to the posterior inferotemporal cortex. It receives information directly from V1, especially related to central space. There is a weak connection to V5. Like V1 it is tuned for colour, and also orientation and spatial frequency, and simple geometric shapes. It is the third cortical area in the ventral stream.

V5, also known as the MT (middle temporal) area, is linked to the magno-cellular pathway from the LGN, It has an important role in perception of motion and guiding the movement of the eye, and transmits to the parietal cortex.

It is a commonly held fallacy that the eyes send an image to the brain, which although upside down and back-to-front, nevertheless manages to sort it out from experience of the real world. As shown above, it is a much more complicated process. The 'image' consists of a series of chemical reactions and electrical impulses firing in specific patterns to various sections of the brain. It is a breathtaking achievement for the brain to interpret these patterns into images of the world around us, and one most of us take for granted.

What causes light to become coloured?

Colour is reflected light which varies in wavelength, and is perceived in different hues of blue, green, yellow and red and mixes of these colours. Light consists of electromagnetic radiation, but with different wavelengths and frequencies. A wavelength is the distance between two consecutive peaks of activity. Wavelengths are measured in nanometres. A nanometre is one billionth of a metre. White light passing through a prism shows the colours of the spectrum. We perceive colour in three dimensions – hue, brightness and saturation. The brightness varies from dark to light. The hue varies continuously around the colour ranges. Saturation varies from rich colours at the periphery of the retina to poor grey tones at the centre. For example red would be at the edge and pink would be at the centre.

The cone photoreceptors in the retina are stimulated by wavelengths. Cones contain types of opsin that are sensitive to various wavelengths of light, which is the basis for colour vision. As previously mentioned, an opsin is a component of the photopigment in the retina. Different cones react to wavelengths that are short, medium or long.

Electromagnetic fields vibrate at right angles to the direction of movement of a wave and at right angles to each other. Electromagnetic radiation consists of small amounts of energy called quanta. Each quantum has a wavelength number. For example a wavelength of 690 nanometres (nm) is the colour red. The human visual system responds to wavelengths of about 400 – 700 nanometres, which is a narrow section of the total range. Each quantum of light energy is called a photon.[4] Figure 19.6 shows the range of wavelengths in metres. To explain the numbers, number 10^{-4} means minus 1 and 4 zeros (˙10,000). The number 10^2 means 1 and 2 zeros (10 x 10 = 100).

There have been various theories attempting to explain how we see different colours. The one that seems to have gained precedence is that proposed by Thomas Young in 1802. He suggested that the eye detects different colours because there

Figure 19.6 Wavelength range (in metres)

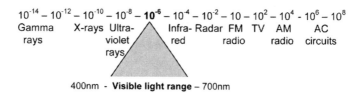

400nm - **Visible light range** – 700nm

are three types of receptors (cones), each one responsive to a single hue. The primary colours he noted were red, blue and green. The short-wave cones are stimulated by blue, the medium-wave cones are stimulated by green and the long-wave cones are responsive to the colour red.

His hypothesis is called the trichromatic (three-colour) theory. Mixing these colours produces the different range of colours along the spectrum. This is not the same process as mixing paint colours, when we are actually mixing pigments, for example red and blue become purple. Colour mixing is when several light sources are connected. If two beams of red and bluish green light are shined on a white screen, the result is a yellow light. A mix of yellow and blue light will produce a white light. The white colour seen on a colour television is made up of tiny pixels of red, blue and green light.[5] Various later studies have confirmed his theory. However it does not explain why we consider yellow to also be one of the main primary colours. Nor does it explain why some colours blend like red and blue, but others do not, such as yellow and blue.

Other scientists have suggested that the three-colour code becomes translated into an opponent-colour system.[6] Their studies found that two types of retinal ganglion cells respond to pairs of primary colours that oppose each other. One pair has red/green and the other has yellow/blue. As previously mentioned the ganglion cells consist of two concentric circles. A cell may be excited by one colour in its outside ring which inhibits the pairing colour at the centre. Figure 19.7 shows how this may work.[7]

Figure 19.7 Colour-sensitivity of receptive fields in ganglion cells

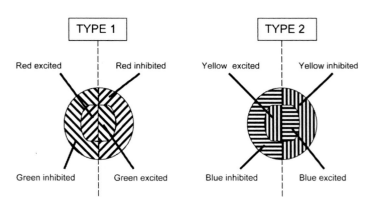

This system explains why we are unable to see a reddish-green mix or a yellow-blue mix as both cannot be excited or inhibited at the same time. It appears that the two systems work in tangent. The three types of cones connect with the two types of ganglion cells and form neuronal circuits with the other retinal cell types: bipolar, amacrine, and horizontal.

What is colour-blindness?

Colour-blindness is the inability to perceive certain differences in colours that most people can distinguish. About 8% of men and 0.5% of women have an element of colour blindness.[8] It can be caused by damage to the retina, optic nerve or the regions of the brain involved in colour processing, such as the parvocellular pathway of the LGN in the thalamus and the V4 area of the visual cortex (see Figures 19.4, 19.5, and 19.6).

However, the most common cause is genetically inherited,

and results from defects in one or more of the three types of cone receptor. The source of the problem lies with genes on the X chromosome. Males have only one X chromosome, but females have two. This means that females are less likely to have the condition, because the genes on one X chromosome will probably be normal and will compensate for the defective gene. This accounts for the much higher proportion of men who are colour-blind.

There are two conditions that confuse the colours red and green. To people with protanopia, both colours look a kind of yellow. It is thought their 'red' cones are filled with 'green' cone opsin. In people with deuteranopia, their 'green' cones are filled with 'red' cone opsin. A much more rare disorder is tritanopia, which affects men and women equally. There is difficulty perceiving short wavelength hues, caused by a lack of 'blue' cones. Their world is coloured green and red from the medium and long wavelengths. A blue sky looks green and yellow appears pink.

Can we trust what we see – illusion and reality?

We perceive our world in a seamless 'wide-screen' aspect, but that is not strictly accurate. Where the optic nerve exits at the back of the eye, it causes a break in the retina where there are no photoreceptor cells (see Figure 19.1). An image entering this area will not be transmitted, and it is called the blind spot. We can carry out an experiment to test this. Look at Figure 19.8. Close your right eye and focus on the small black circle on the right with your left eye. Hold the page about 12 cm away. The black square will disappear. You may have to move a little nearer to the page to get the full effect.

Figure 19.8 Blind spot experiment 1

■   ●

The brain compensates for the blind spot by filling in the gap, based on our experience of the world. Figure 19.9 shows an even clearer example of this. Close your left eye and focus on the X with your right eye. The gap between the two solid oblongs will disappear.

Figure 19.9 Blind spot experiment 2

Some people who have suffered brain damage may display a loss of visual perception in specific ways. This condition is called agnosia. A study was carried out with a patient who had a reduced blood supply to the temporal cortex at the back of his brain from a blocked cerebral artery.[9] Although he said he was unable to recognise objects shown to him, he could describe them well when answering questions, or by touch or sound. He could copy a drawing, which showed that he could see, but was unable to say what it was.

Blindsight is a condition resulting from damage in one hemisphere to the primary visual cortex (see Figure 19.5). Patients say they are unable to see objects in the area covered by this part of the brain, but react as though they can see. For example, in one study, a patient who had accidentally inhaled carbon monoxide gas was thereafter unable to recognise faces or name objects.[10] Although she believed she did not know which way a slit in a piece of card was orientated, she could accurately put her hand through the slit, at the right angle. Therefore she could see, but did not realise it.

Finally we turn to visual illusions and ambiguous figures.

Figure 19.10 Visual illusions

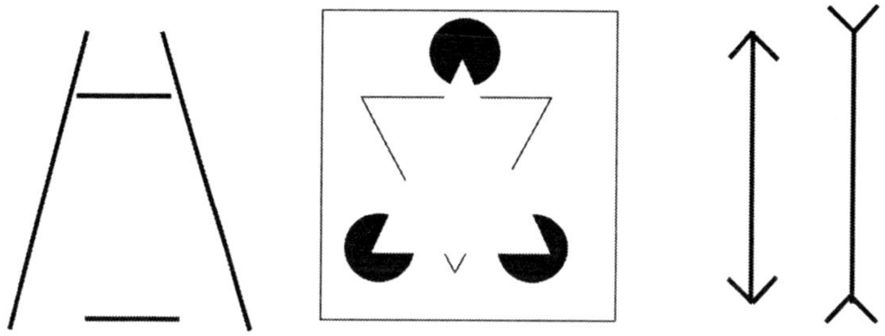

(A)   (B)   (C)

Figure 19.10 (A) shows a figure that looks like railway lines. The bottom line looks shorter than the top line, but in fact they are both the same size. The brain is accustomed to interpreting perspective where things that are in the distance are actually larger than they appear. Figure 19.10 (B) shows the Kanizsa triangles.[11] Most people will see an image of a white triangle. The image appears to be above and partly blocking another triangle and three circles. In fact the drawings are of a single triangle, which has a gap on each side, and the circles have a chunk missing. It is an illusion. In Figure 19.10 (C) the arrow on the left appears to be shorter than the one on the right. Actually, they are both the same size. It is the angle of the arrow points at each end of the lines that alters the perception of size.

These ambiguous figures are caused by the brain being undecided about which interpretation is correct, and without any other corroborating stimuli, reverses from one interpretation to an opposing one. Figure 19.11 (A) shows a cube, which can be seen as complete. Stare at it for a little while, and the perspective of the cube changes, so that the shaded part is seen as the ceiling of a two-sided room. In Figure 19.11 (B) the image of a folded card will change from an outward to an inward fold.

The upshot of all this is that we cannot always believe what we see.

Figure 19.11 Ambiguous figures

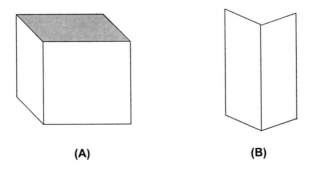

(A)   (B)

Why do we need to make eye-contact?

It has been estimated that people will look at each other during a conversation about 60% of the time, half of which maintains eye-contact.[12] Of the remaining 30%, their eyes will move quickly between the eyes and mouth of the person with whom they are talking. Human babies establish eye-contact with their mother by the fourth week. It appears to have significant social implications, especially related to seeking and paying attention during social interactions.

We make eye-contact when greeting or saying goodbye. It also is an important factor in sexual attraction. Women will try to enhance the size of their eyes by using make-up to focus attention on that area of the face.

People with psychiatric disorders such as autism, schizophrenia and depression are unable or unwilling to hold the gaze in a normal way. During conversations, they may stare in an abnormal way, or avert their eyes. It is an indication that something is amiss. It appears that extraverts gaze more than introverts, women more than men, and children and adults more than adolescents.

Blinking sometimes increases during times of acute anxiety or embarrassment. This does not seem to be related to keeping the eyes clean but may be an unconscious attempt to avoid eye-contact throughout a stressful period.

Does eating carrots improve eyesight?

Carrots are a good source of vitamin A, beta-carotene and

potassium – all essential for healthy eyesight, skin, growth and resisting infection. However, it is a myth that eating carrots will improve eyesight.

One theory how the myth started seems to be during World War 2. The War Ministry were afraid the enemy would discover that the increased success of the Royal Air Force intercepting aircraft was due to the discovery of night radar. So they spread the misinformation story that it was due to eating carrots, which improved vision. Apparently one eagle-eyed ace pilot was very fond of them and the media story spread – and a myth was born. However, beta-carotene if taken in high doses can reduce the risk of cataract and macular degeneration.
A deficiency of vitamin A also causes the condition known as night-blindness.

Large doses of vitamin A can be toxic and too much beta-carotene results in carotenemia, which turns the skin yellow or orange. So eating carrots is definitely good for you – but not too many.

Disorders affecting the eye

Long- and short-sightedness – As mentioned earlier in this chapter, the focus of the image is controlled by changes in the shape of the lens. Long-sightedness is also known as hyperopia or far-sightedness. People with this condition are unable to focus on close objects. It is caused by the eyeball being too short or the cornea too flat. Light entering the eye focuses behind the retina instead of on it, and close objects become blurred.

People with short-sight are unable to focus on distant objects, which appear blurred. The alternative names for this condition are myopia or near-sightedness. In this case, the eyeball is too long and light focuses in front of the retina. Both conditions are easily rectified by wearing spectacles or contact lenses.

Astigmatism is often associated with long- and short-sightedness. In this situation, the shape of the cornea is elongated, rather like a rugby ball, instead of circular. Light entering the eye is bent more sharply in one direction of the cornea, resulting in an inability to focus correctly.

Cataract – This is not, as is popularly understood, a skin growing over the eye, but a clouding of the lens. Light cannot pass through, and results in dim or blurred vision. It is a natural consequence of the ageing process. It can also be caused from an injury, medication or diabetes.

Macular degeneration – Delicate cells in the macula, which is the small central part of the retina, become damaged, most commonly caused by age. There are two types. The dry form develops gradually, and causes mild loss of vision. In a minority of cases, the wet form, which is much rarer, causes a higher risk of loss of vision.

A similar condition can be inherited called macular dystrophy. It only affects the central part of vision, which can be distorted or sensitive to light. There may also be a difficulty distinguishing colours. Peripheral sight is unaffected. Although the cause is not known, there are known high-risk groups. These include smokers, those on a high-fat diet and people with a high exposure to the sun.

Glaucoma – The loss of retinal ganglion cells, caused by damage to the optic nerve creates this condition. It results in increased pressure in the eye, and is inherited. People over the age of 40 are at greater risk, as well as diabetics or the extremely short-sighted.

Detached retina – In this condition, the retina becomes separated from its underlying tissue, caused by a tear in the retina, and fluid accumulates. It is often caused by an injury or accident. The symptoms are a shadow across the visual field of the eye, bright flashes of light or dark spots (floaters). Prompt laser or freezing treatment minimises damage. If the retina is completely detached, an operation can be carried out to repair the tear.

REFERENCES

Rosenzweig, M. R., Breedlove, S.M., Watson, N.V., (2005) *Biological Psychology*, Fourth Edition, Sinauer Associates, Inc., p.293

[2] Rosenzweig, M.R., Breedlove, S.M., Watson, N.V., (2005) op. cit. p.291

[3] Milner, A.D., and Goodale, M., (1995) *The Visual Brain in Action*, Oxford University Press, Oxford

[4] Rosenzweig, M.R., Breedlove, S.M., Watson, N.V., (2005) op. cit. p. 288

[5] Carlson, N.R., (1994) *Physiology of Behavior*, Fifth Edition, Allyn and Bacon, p. 153

[6] Daw, N.W., (1968) Colour-coded ganglion cells in the goldfish retina: Extension of their receptive fields by means of new stimuli. *Journal of Physiology, London, 197*, pp.567-592 and Gouras, P., (1968), Identification of cone mechanisms in monkey ganglion cells. *Journal of Physiology, London, 199* pp. 553-538

[7] Carlson, N.R, (1994) *Physiology of Behavior*, op. cit., p. 155

[8] Rosenzweig, M.R., Breedlove, S.M., Watson, N.V., (2005) op. cit. p. 288

[9] Riddoch, M.J., and Humphries, G.W., (1987). A case of integrative visual agnosia, *Brain, 110*, pp. 1431-62

[10] Goodale, M.A., Milner, A.D., Jakobson, L.S., and Carey, D.P., (1991). A

neurological dissociation between perceiving objects and grasping them. *Nature, 349*, pp. 154-156

[11] Kanizsa, G., (1955), "Margini quasi-percettivi in campi con stimolazione omogenea" *Rivista di Psicologia* **49** (1):07-30

[12] Gregory, R.L., (Ed) and Zangwill, O.L. (1987) The Oxford Companion to the Mind, Eye-contact, Oxford University Press, p. 247